Tensor Calculus, Relativity, and Cosmology

A First Course

Tensor Calculus, Relativity, and Cosmology

A First Course

M. Dalarsson
Ericsson Research and Development
Stockholm, Sweden

and

N. Dalarsson
Royal Institute of Technology
Stockholm, Sweden

Amsterdam • Boston • Heidelberg • London • New York • Oxford
Paris • San Diego • San Francisco • Singapore • Sydney • Tokyo

ELSEVIER
ACADEMIC
PRESS

Acquisitions Editor: Jeremy Hayhurst
Editorial Assistant: Desiree Marr
Publishing Services Manager: Andre Cuello
Project Manager: Justin Palmeiro
Marketing Manager: Linda Beattie
Cover Design: Diana Coe
Composition: Newgen Imaging Systems (P) Ltd.
Cover Printer: Phoenix
Interior Printer: Maple-Vail

Elsevier Academic Press
30 Corporate Drive, Suite 400, Burlington, MA 01803, USA
525 B Street, Suite 1900, San Diego, California 92101-4495, USA
84 Theobald's Road, London WC1X 8RR, UK

This book is printed on acid-free paper

Library of Congress Cataloging-in-Publication Data

Application Submitted.

British Library Cataloguing in Publication Data
A catalogue record for this book is available from the British Library

ISBN: 0-12-200681-X

For all information on all Elsevier Academic Press Publications
visit our Web site at www.books.elsevier.com

Printed in the United States of America
05 06 07 08 09 10 9 8 7 6 5 4 3 2 1

Working together to grow
libraries in developing countries

www.elsevier.com | www.bookaid.org | www.sabre.org

ELSEVIER   BOOK AID
            International   Sabre Foundation

Preface

The subjects of relativity, gravitation, and cosmology are frequently discussed areas of modern physics among the broad audience. Numerous popular books on these subjects are available for the general reader with no background in physics. At the same time a number of textbooks and monographs for advanced graduate students and young researchers in the area is available as well.

However, intermediate books, allowing senior undergraduate students or junior PhD students to enter this exciting area of physics in a smooth and pedagogical way, are quite rare. It is generally assumed that new concepts of both physics and mathematics involved in the subject are too complex to allow a simple and pedagogical introduction.

The only way to master the subject seems to be a hard and time-consuming effort to put together the intelligible pieces of advanced textbooks into an understandable set of notes, as experienced by ourselves and many of our colleagues as well. This approach has the advantage of making young students with the capability and perseverance to go through such a learning process very well trained for future research tasks. However, at the same time, it severely limits the number of people who ever really master the subject.

This book is an attempt to bridge the gap between a regular university curriculum, consisting typically of courses in calculus and general physics, and the more advanced books in tensor calculus, relativity, and cosmology. The book has evolved from a set of lecture notes originally compiled by one of the authors, M. Dalarsson, but has been improved and completed with a few exciting new topics over the past 10 years.

It is the intention of the book to give to the readers a high level of detail in derivations of all equations and results. The more lengthy and tedious algebraic manipulations are in general outlined in such detail that they can be followed by an interested senior undergraduate student or a junior PhD student with very little or no risk of ever getting lost.

It is our experience that a common showstopper for a young university student trying to master a subject are phrases in the literature claiming that something can be derived from something else by some "straightforward although somewhat tedious algebra." If a student cannot readily reproduce such a "straightforward" algebra, which most often is the case, the usual reaction under the time pressure of the studies is to accept the claim as a fact. And from that point, throughout the rest of the course, the deeper understanding of the subject is lost.

There are a number of advanced books on relativity, gravitation, and cosmology, and we have benefited from some of those as well as from unpublished notes produced by some of our distinguished colleagues. Some of these sources are listed in the bibliography at the end of the book, as well as a few other books that may be recommended as suitable further reading. However, in an introductory book such as this one, it is possible neither to include an extensive list of all original references and major textbooks and monographs on the subject nor to mention all the people who have contributed to our understanding of this exciting subject.

We hope that our readers will find that we have, at least partly, fulfilled the objective of bridging the gap between the regular university curriculum and the more advanced literature on this exciting subject, and that they will enjoy reading this book as much as we did writing it.

M. Dalarsson and N. Dalarsson
Stockholm, Sweden, January 2003

Contents

1   Introduction   **1**

Part I   Tensor Algebra   **3**

2   Notation and Systems of Numbers   **5**
  2.1   Introduction and Basic Concepts   5
  2.2   Symmetric and Antisymmetric Systems   7
  2.3   Operations with Systems   8
      2.3.1   Addition and Subtraction of Systems 8
      2.3.2   Direct Product of Systems   8
      2.3.3   Contraction of Systems   9
      2.3.4   Composition of Systems   9
  2.4   Summation Convention   10
  2.5   Unit Symmetric and Antisymmetric Systems   11

3   Vector Spaces   **15**
  3.1   Introduction and Basic Concepts   15
  3.2   Definition of a Vector Space   16
  3.3   The Euclidean Metric Space   18
  3.4   The Riemannian Spaces   18

4   Definitions of Tensors   **23**
  4.1   Transformations of Variables   23
  4.2   Contravariant Vectors   24
  4.3   Covariant Vectors   24
  4.4   Invariants (Scalars)   24
  4.5   Contravariant Tensors   25
  4.6   Covariant Tensors   26

4.7   Mixed Tensors   26
4.8   Symmetry Properties of Tensors   27
4.9   Symmetric and Antisymmetric Parts of Tensors   28
4.10 Tensor Character of Systems   30

5   Relative Tensors   **33**
5.1   Introduction and Definitions   33
5.2   Unit Antisymmetric Tensors   34
5.3   Vector Product in Three Dimensions   36
5.4   Mixed Product in Three Dimensions   38
5.5   Orthogonal Coordinate Transformations   39
      5.5.1   Rotations of Descartes Coordinates   39
      5.5.2   Translations of Descartes Coordinates   41
      5.5.3   Inversions of Descartes Coordinates   41
      5.5.4   Axial Vectors and Pseudoscalars in Descartes
              Coordinates   42

6   The Metric Tensor   **43**
6.1   Introduction and Definitions   43
6.2   Associated Vectors and Tensors   46
6.3   Arc Length of Curves: Unit Vectors   48
6.4   Angles between Vectors   49
6.5   Schwarz Inequality   51
6.6   Orthogonal and Physical Vector Coordinates   52

7   Tensors as Linear Operators   **55**

Part II   Tensor Analysis   **59**

8   Tensor Derivatives   **61**
8.1   Differentials of Tensors   61
      8.1.1   Differentials of Contravariant Vectors   64
      8.1.2   Differentials of Covariant Vectors   64
8.2   Covariant Derivatives   65
      8.2.1   Covariant Derivatives of Vectors   65
      8.2.2   Covariant Derivatives of Tensors   66
8.3   Properties of Covariant Derivatives   67
8.4   Absolute Derivatives of Tensors   69

9   Christoffel Symbols   **71**
9.1   Properties of Christoffel Symbols   71
9.2   Relation to the Metric Tensor   74

10   Differential Operators   **79**
  10.1   The Hamiltonian ∇-Operator   79
  10.2   Gradient of Scalars   79
  10.3   Divergence of Vectors and Tensors   80
  10.4   Curl of Vectors   82
  10.5   Laplacian of Scalars and Tensors   83
  10.6   Integral Theorems for Tensor Fields   85
         10.6.1   Stokes Theorem   85
         10.6.2   Gauss Theorem   86

11   Geodesic Lines   **89**
  11.1   Lagrange Equations   89
  11.2   Geodesic Equations   92

12   The Curvature Tensor   **97**
  12.1   Definition of the Curvature Tensor   97
  12.2   Properties of the Curvature Tensor   100
  12.3   Commutator of Covariant Derivatives   103
  12.4   Ricci Tensor and Scalar   104
  12.5   Curvature Tensor Components   105

Part III   Special Theory of Relativity   **109**

13   Relativistic Kinematics   **111**
  13.1   The Principle of Relativity   111
  13.2   Invariance of the Speed of Light   112
  13.3   The Interval between Events   112
  13.4   Lorentz Transformations   116
  13.5   Velocity and Acceleration Vectors   119

14   Relativistic Dynamics   **123**
  14.1   Lagrange Equations   123
  14.2   Energy–Momentum Vector   125
         14.2.1   Introduction and Definitions   125
         14.2.2   Transformations of Energy–Momentum   128
         14.2.3   Conservation of Energy–Momentum   130
  14.3   Angular Momentum Tensor   131

15   Electromagnetic Fields   **135**
  15.1   Electromagnetic Field Tensor   135
  15.2   Gauge Invariance   140
  15.3   Lorentz Transformations and Invariants   142

16   Electromagnetic Field Equations   **147**
   16.1   Electromagnetic Current Vector   147
   16.2   Maxwell Equations   149
   16.3   Electromagnetic Potentials   154
   16.4   Energy–Momentum Tensor   155

Part IV   General Theory of Relativity   **163**

17   Gravitational Fields   **165**
   17.1   Introduction   165
   17.2   Time Intervals and Distances   167
   17.3   Particle Dynamics   169
   17.4   Electromagnetic Field Equations   173

18   Gravitational Field Equations   **177**
   18.1   The Action Integral   177
   18.2   Action for Matter Fields   182
   18.3   Einstein Field Equations   188

19   Solutions of Field Equations   **193**
   19.1   The Newton Law   193
   19.2   The Schwarzschild Solution   195

20   Applications of the Schwarzschild Metric   **207**
   20.1   The Perihelion Advance   207
   20.2   Black Holes   215

Part V   Elements of Cosmology   **223**

21   The Robertson–Walker Metric   **225**
   21.1   Introduction and Basic Observations   225
   21.2   Metric Definition and Properties   227
   21.3   The Hubble Law   234
   21.4   The Cosmological Red Shifts   235

22   Cosmic Dynamics   **239**
   22.1   The Einstein Tensor   239
   22.2   The Friedmann Equations   250

23   Nonstatic Models of the Universe   **253**
   23.1   Solutions of the Friedmann Equations   253
        23.1.1   The Flat Model ($k = 0$)   255

23.1.2 The Closed Model ($k = 1$) 255
23.1.3 The Open Model ($k = -1$) 257
23.2 Closed or Open Universe 258
23.3 Newtonian Cosmology 260

24 Quantum Cosmology **265**
24.1 Introduction 265
24.2 The Wheeler–DeWitt Equation 266
24.3 The Wave Function of the Universe 270

Bibliograhy **275**
Index **277**

Chapter 1

Introduction

The tensor calculus is a mathematical discipline of relatively recent origin. It is fair to say that, with few exceptions, the tensor calculus was developed during the twentieth century. It is also an area of mathematics that was developed for an immediate practical use in the theory of relativity, with which it is strongly interrelated. Later, however, the tensor calculus has proven to be useful in other areas of physics and engineering such as classical mechanics of particles and continuous media, differential geometry, electrodynamics, quantum mechanics, solid-state physics, and quantum field theory. Recently, it has been used even in electric circuit theory and some other purely engineering disciplines.

In the early twentieth century, at the same time when the tensor calculus was developed, a number of major breakthroughs in modern science were made. In 1905 the special theory of relativity was formulated, then in 1915 the general theory of relativity was developed, and in 1925 quantum mechanics took its present form. In the years to come quantum mechanics and the special theory of relativity were combined to develop the relativistic quantum field theory, which gives at least a partial explanation of the three fundamental forces of nature (strong, electromagnetic, and weak).

The remaining known fundamental force of nature, the force of gravity, is different from the other three fundamental forces. Although very weak on the small scale, gravity dominates the other three forces over cosmic distances. This dominance, due to gravity being a long-range force that cannot be screened, makes it the only available foundation for any cosmology. The other three fundamental forces are explained through particle interactions in the flat space-time of special relativity. However, gravity does not allow for such an explanation. In order to explain gravity, Einstein had to connect it with the geometry of space-time and formulate a relativistic theory of gravitation. For a long time, general relativity was separate from the other parts of physics, partly because of the mathematical framework of the theory (tensor calculus), which was not extensively used in any other discipline during that time.

The tensor calculus today is used in a number of other disciplines as well, and its extension to other areas of physics and engineering is a result of the simplification of the mathematical notation and in particular the possibility of natural extension of the equations to the relativistic case.

Today, physics and astronomy have joined forces to form the discipline called relativistic astrophysics. The major advances in cosmology, including the first attempts to formulate quantum cosmology, also increase the importance of general relativity. Finally, a number of attempts have been made to unify gravity with the other three fundamental forces of nature, thus introducing the tensor calculus and Riemannian geometry to the new exciting areas of physics such as the theory of superstrings.

In the first two parts of the book a pedagogical introduction to the tensor calculus is covered. Thereafter, an introduction to the special and general theories of relativity is presented. Finally an introduction to the modern theory of cosmology is discussed.

Part I

Tensor Algebra

► Chapter 2

Notation and Systems of Numbers

$\boxed{2.1}$ Introduction and Basic Concepts

In order to get acquainted with the basic notation and concepts of the tensor calculus, it is convenient to use some well known concepts from linear algebra. The collection of N elements of a column matrix is often denoted by subscripts as $x_1, x_2, \ldots, x_N$. Using a lower index $i = 1, 2, \ldots, N$, we can introduce the following short-hand notation:

$$x_i \quad (i = 1, 2, \ldots, N). \tag{2.1}$$

Sometimes, the same collection of N elements is denoted by corresponding superscripts as $x^1, x^2, \ldots, x^N$. Using here an upper index $i = 1, 2, \ldots, N$, we can also introduce the following short-hand notation:

$$x^i \quad (i = 1, 2, \ldots, N). \tag{2.2}$$

In general the choice of a lower or an upper index to denote the collection of N elements of a column matrix is fully arbitrary. However, it will be shown later that in the tensor calculus lower and upper indices are used to denote mathematical objects of different natures. Both types of indices are therefore essential for the development of tensor calculus as a mathematical discipline. In the definition (2.2) it should be noted that i is an upper index and not a power of x. Whenever there is a risk of confusion of an upper index and a power, such as when we want to write a square of x^i, we will use parentheses as follows:

$$x^i \cdot x^i = (x^i)^2 \quad (i = 1, 2, \ldots, N). \tag{2.3}$$

A collection of numbers, defined by just one (upper or lower) index, will be called a first-order system or a simple system. The individual elements of such a system will be called the elements or coordinates of the system. The introduction of the lower and upper indices provides a device to highlight the different nature of different first-order systems with equal numbers of elements. Consider, for example, the following linear form:

$$ax + by + cz. \tag{2.4}$$

Introducing the labels $a_i = \{a, b, c\}$ and $x^i = \{x, y, z\}$, the expression (2.4) can be written as

$$a_1 x^1 + a_2 x^2 + a_3 x^3 = \sum_{i=1}^{3} a_i x^i, \tag{2.5}$$

indicating the different nature of the two first-order systems. In order to emphasize the advantage of the proposed notation, let us consider a bilinear form created using two first-order systems x^i and y^i $(i = 1, 2, 3)$.

$$a_{11} x^1 y^1 + a_{12} x^1 y^2 + a_{13} x^1 y^3 + a_{21} x^2 y^1 + a_{22} x^2 y^2 + a_{23} x^2 y^3$$

$$+ a_{31} x^3 y^1 + a_{32} x^3 y^2 + a_{33} x^3 y^3 = \sum_{i=1}^{3} \sum_{j=1}^{3} a_{ij} x^i y^j \tag{2.6}$$

Here we see that the short-hand notation on the right-hand side of Eq. (2.6) is quite compact. The system of parameters of the bilinear form

$$a_{ij} \quad (i, j = 1, 2, 3), \tag{2.7}$$

is labeled by two lower indices. This system has nine elements and they can be represented by the following 3×3 square matrix:

$$\begin{bmatrix} a_{11} & a_{12} & a_{13} \\ a_{21} & a_{22} & a_{23} \\ a_{31} & a_{32} & a_{33} \end{bmatrix}. \tag{2.8}$$

A system of quantities determined by two indices is called a second-order system.

Depending on whether the indices of a second-order system are upper or lower, there are three types of second-order systems:

$$a_{ij}, \quad a_i^j, \quad a^{ij} \quad (i, j = 1, 2, \ldots, N). \tag{2.9}$$

A second-order system in N dimensions has N^2 elements. In a similar way we can define the third-order systems, which may be of one of four different types:

$$a_{ijk}, \quad a^i_{jk}, \quad a^{ij}_k, \quad a^{ijk} \quad (i,j = 1, 2, \ldots, N). \tag{2.10}$$

The most general system of order K is denoted by

$$a_{i_1, i_2, \ldots, i_K} \quad (i_1, i_2, \ldots, i_K = 1, 2, \ldots, N), \tag{2.11}$$

and, depending on the position of the indices, it may be of one of several different types. The Kth-order system in N dimensions has N^K elements.

2.2 Symmetric and Antisymmetric Systems

Let us consider a second-order system in three dimensions

$$a_{ij} \quad (i,j = 1, 2, 3). \tag{2.12}$$

The system (2.12) is called a symmetric system with respect to the two lower indices if the elements of the system satisfy the equality

$$a_{ij} = a_{ji} \quad (i,j = 1, 2, 3). \tag{2.13}$$

Similarly, the system (2.12) is called an antisymmetric system with respect to the two lower indices if the elements of the system satisfy the equality

$$a_{ij} = -a_{ji} \quad (i,j = 1, 2, 3). \tag{2.14}$$

The equality (2.14) indicates that an antisymmetric second-order system in three dimensions has only three independent components and that all the diagonal elements are equal to zero:

$$a_{JJ} = 0 \quad (J = 1, 2, 3). \tag{2.15}$$

Thus it is possible to represent an antisymmetric second-order system in three dimensions by the following 3×3 matrix:

$$\begin{bmatrix} 0 & a_{12} & a_{13} \\ -a_{12} & 0 & a_{23} \\ -a_{13} & -a_{23} & 0 \end{bmatrix}. \tag{2.16}$$

In general, a system of an arbitrary order and type will be symmetric with respect to two of its indices (both upper or both lower), if the corresponding elements remain unchanged upon interchange of these two indices. The system will be totally symmetric with respect to all upper (lower) indices, if an interchange of any two upper (lower) indices leaves the corresponding system elements unchanged. Elements of a totally

symmetric third-order system with all three lower indices satisfy the
equality

$$a_{ijk} = a_{ikj} = a_{jki} = a_{jik} = a_{kij} = a_{kji}. \tag{2.17}$$

Analogously to the above, a system of an arbitrary order and type will be
antisymmetric with respect to the two of the indices (both upper or both
lower), if the corresponding elements change signs upon interchange of
these two indices. The system will be totally antisymmetric with respect to
all upper (lower) indices, if an interchange of any two upper (lower) indices
changes signs of the corresponding system elements. Elements of a totally
antisymmetric third-order system with all three lower indices satisfy the
equality

$$a_{ijk} = -a_{ikj} = a_{jki} = -a_{jik} = a_{kij} = -a_{kji}. \tag{2.18}$$

2.3 Operations with Systems

Under certain conditions, it is possible to perform a number of algebraic
operations with systems. The definition of these operations depends on the
order and type of the systems.

2.3.1 Addition and Subtraction of Systems

The addition and subtraction of systems can be performed only with
systems of the same order and same type. The addition (subtraction) of
systems is performed in such a way that each element of one system is
added (subtracted) to (from) the corresponding element of the other system
(the one with the same indices in the same order). For example, the systems
A_{km}^{ij} and B_{km}^{ij} can be added since they are of the same order and of the same
type. The sum of these two systems is given by

$$D_{km}^{ij} = A_{km}^{ij} + B_{km}^{ij}, \tag{2.19}$$

and it is a system of the same order and type as the two original systems.
This definition is easily extended to addition and subtraction of an arbitrary
number of systems.

2.3.2 Direct Product of Systems

A system obtained by multiplying each element of one system by each
element of another system, regardless of their order and type, is called

a direct product or just a product of these two systems. Thus, for example, a product of two first-order systems a^i and b^i is a second-order system

$$c^{ij} = a^i b^j. \tag{2.20}$$

For $i, j = 1, 2, 3$, this operation can be written in the following matrix form:

$$[c^{ij}] = \begin{bmatrix} a^1 \\ a^2 \\ a^3 \end{bmatrix} \begin{bmatrix} b^1 & b^2 & b^3 \end{bmatrix} = \begin{bmatrix} a^1 b^1 & a^1 b^2 & a^1 b^3 \\ a^2 b^1 & a^2 b^2 & a^2 b^3 \\ a^3 b^1 & a^3 b^2 & a^3 b^3 \end{bmatrix}. \tag{2.21}$$

In general, the set of upper (lower) indices of a system, created as a product of several other systems, is a collection of all upper (lower) indices of all of the constituent systems. For example, we have

$$D_{jln}^{ikm} = A_j^i B_l^k C_n^m. \tag{2.22}$$

2.3.3 Contraction of Systems

This operation is applicable to systems with at least one pair of indices of opposite type, i.e., at least one upper index and one lower index. The actual pair of indices of opposite type is then made equal to each other and a sum over that common index is performed. Thus, for example, by contraction of a third-order system a_k^{ij}, we obtain

$$a^i = \sum_{j=1}^{N} a_j^{ij} = a_1^{i1} + a_2^{i2} + \cdots + a_N^{iN}. \tag{2.23}$$

The contraction of a system of order k gives a system of order $k - 2$, which is easily seen from the example (2.23). The contraction of a mixed second-order system b_j^i gives a zeroth-order system

$$b = \sum_{j=1}^{N} b_j^j = b_1^1 + b_2^2 + \cdots + b_N^N, \tag{2.24}$$

which is equal to the trace of the matrix $[b_j^i]$.

2.3.4 Composition of Systems

The composition of systems is a complex operation consisting of a product of two systems and a contraction with respect to at least one of the indices of opposite type from each of the systems. The product of the two first-order systems a^k and b_j is a mixed second-order system $a^k b_j$. By contraction of this system, we obtain the composition of the systems a^k and b_j in the form

$$c = \sum_{j=1}^{N} a^j b_j = a^1 b_1 + a^2 b_2 + \cdots + a^N b_N. \tag{2.25}$$

For $N = 3$, the result (2.25) can be written in the following matrix form:

$$\sum_{j=1}^{N} a^j b_j = \begin{bmatrix} a^1 & a^2 & a^3 \end{bmatrix} \begin{bmatrix} b^1 \\ b^2 \\ b^3 \end{bmatrix}. \tag{2.26}$$

2.4   Summation Convention

At the very beginning of the development of the general theory of relativity, in order to simplify derivations of various results and expressions in the tensor calculus, the summation convention over the repeated indices was introduced. According to this convention, when the same index in an expression appears twice, it is understood that the summation over that index is performed and no summation sign is needed. Thus, for example, we may write

$$\sum_{j=1}^{N} a_j x^j = a_j x^j, \tag{2.27}$$

$$\sum_{j=1}^{N} \sum_{k=1}^{N} a_{jk} x^j x^k = a_{jk} x^j x^k, \tag{2.28}$$

$$\sum_{j=1}^{N} \sum_{k=1}^{N} D_{jk}^{mn} a^{jk} = D_{jk}^{mn} a^{jk}. \tag{2.29}$$

The repeated indices, over which a summation is understood, are usually called *dummy indices*. When using the summation convention, the following rules should be kept in mind:

1. It is required to know exactly which range of values all indices can take. If nothing else is specified, it is assumed that all indices in one expression or equation cover the same range of integers.

2. When it is required to represent any of the three diagonal elements a_1^1, a_2^2 or a_3^3, in order to avoid confusion with the summation convention, $a_M^M (M = 1, 2, 3)$ could be used instead of $a_m^m, (m = 1, 2, 3)$. In such a case the capital letters are only used for this purpose and are otherwise not used as indices.

3. In order to avoid confusion, whenever there are two or more pairs of dummy indices in the same expression, they should always be denoted by different letters and never by the same letters.

2.5 Unit Symmetric and Antisymmetric Systems

The unit symmetric system, called the δ-symbol, is the symmetric second-order system, defined as follows:

$$\delta_j^i = \begin{cases} 1, & \text{for } i = j \\ 0, & \text{for } i \neq j \end{cases}. \qquad (2.30)$$

Thus the δ-symbol δ_j^i for $(i, j = 1, 2, 3)$ is a system of nine elements such that the diagonal elements, with indices i and j equal to each other, are equal to unity while the off-diagonal elements are equal to zero. The matrix of this system is the following:

$$\left[\delta_j^i \right] = \begin{bmatrix} 1 & 0 & 0 \\ 0 & 1 & 0 \\ 0 & 0 & 1 \end{bmatrix}. \qquad (2.31)$$

The δ-symbol is often called the substitution operator, since by composition with the δ-symbol it is possible to change the index label of any system. For example, it is possible to write

$$\delta_j^i a_i = \delta_j^1 a_1 + \delta_j^2 a_2 + \delta_j^3 a_3 = a_j. \qquad (2.32)$$

The validity of the result (2.32) is obvious from the definition of the δ-symbol (2.30). Since $\delta_M^M = 1$ for $M = 1, 2, 3$, the trace of the δ-symbol in three dimensions is given by

$$\delta_j^j = \delta_1^1 + \delta_2^2 + \delta_3^3 = 3. \qquad (2.33)$$

The use of the δ-symbol may be illustrated by the following example. If a system of three independent coordinates is given by $x^i = \{x, y, z\}$, these coordinates, by definition, satisfy

$$\frac{\partial x^i}{\partial x^j} = \begin{cases} 1, & \text{for } i = j \\ 0, & \text{for } i \neq j \end{cases}. \tag{2.34}$$

The equation (2.35) can be simplified using the δ-symbol as follows:

$$\frac{\partial x^i}{\partial x^j} = \delta_j^i. \tag{2.35}$$

Another important system is the unit antisymmetric system, that is, the totally antisymmetric third-order system in three dimensions, called the e-system. It is denoted by e^{ijk} or e_{ijk} and is defined for $(i,j,k = 1,2,3)$ in the following way:

$$e^{ijk} = \begin{cases} +1, & \text{if } ijk \text{ is an even permutation of 123} \\ -1, & \text{if } ijk \text{ is an odd permutation of 123} \\ 0, & \text{if } i = j, i = k, j = k \text{ or } i = j = k \end{cases}. \tag{2.36}$$

Let us now write down all $3! = 6$ permutations of 123 in order:

$$\begin{array}{ll} (123) & \text{zeroth permutation} \\ (132) & \text{first permutation} \\ (312) & \text{second permutation} \\ (321) & \text{third permutation} \\ (231) & \text{fourth permutation} \\ (213) & \text{fifth permutation} \end{array} \tag{2.37}$$

The three cyclic permutations (123), (312), and (231) are even permutations and the corresponding e-symbol elements are given by

$$e^{123} = e^{312} = e^{231} = 1. \tag{2.38}$$

Similarly, for the odd permutations the e-symbol elements are given by

$$e^{132} = e^{321} = e^{213} = -1, \tag{2.39}$$

and all other elements of the e-symbol are equal to zero. This indicates that the e-symbol has only six nonzero elements out of a total of 27 elements. Since the e-system in three dimensions is a third-order system, it is not possible to represent it by a two-dimensional matrix, but a three-dimensional scheme is required. However, for better visibility a following schematic representation is also possible:

jk	11	12	13	21	22	23	31	32	33
$i=1$	0	0	0	0	0	1	0	-1	0
$i=2$	0	0	-1	0	0	0	1	0	0
$i=3$	0	1	0	-1	0	0	0	0	0

$$[e^{ijk}] = \qquad\qquad\qquad\qquad\qquad\qquad\qquad\qquad (2.40)$$

The most general Nth-order e-symbol is defined as follows:

$$e^{i_1 i_2 \ldots i_N} = \begin{cases} +1, & i_1 i_2 \ldots i_N \text{ an even permutation of } 12 \ldots N \\ -1, & i_1 i_2 \ldots i_N \text{ an odd permutation of } 12 \ldots N \\ 0, & \text{any pair of indices equal to each other} \end{cases} \quad (2.41)$$

Using the e-symbols it is possible to write down the expression for the determinant of a matrix $[a_j^i]$, for $(i,j = 1,2,3)$, as follows:

$$a = \det\left[a_j^i\right] = e^{kmn} a_k^1 a_m^2 a_n^3. \tag{2.42}$$

In order to verify that the expression (2.42) is indeed equal to the determinant of the matrix $[a_j^i]$, we use the expression for the determinant of a 3×3 matrix as follows:

$$a = \det \begin{bmatrix} a_1^1 & a_2^1 & a_3^1 \\ a_1^2 & a_2^2 & a_3^2 \\ a_1^3 & a_2^3 & a_3^3 \end{bmatrix} \tag{2.43}$$

$$a = a_1^1(a_2^2 a_3^3 - a_2^3 a_3^2) - a_2^1(a_1^2 a_3^3 - a_1^3 a_3^2) + a_3^1(a_1^2 a_2^3 - a_1^3 a_2^2)$$
$$= a_1^1 a_2^2 a_3^3 - a_1^1 a_2^3 a_3^2 + a_3^1 a_1^2 a_2^3 - a_3^1 a_2^2 a_1^3 + a_2^1 a_3^2 a_1^3 - a_2^1 a_1^2 a_3^3 \tag{2.44}$$

From the definition of the e-symbol (2.36) we see that the result (2.44) can be written in the form

$$a = e^{kmn} a_k^1 a_m^2 a_n^3, \tag{2.45}$$

which is identical to (2.42).

The determinant of a matrix $[a_j^i]$, in a general case when $(i,j = 1,2,\ldots,N)$, has the form

$$a = e^{i_1 i_2 \ldots i_N} a_{i_1}^1 a_{i_2}^2 \ldots a_{i_N}^N. \tag{2.46}$$

The definition (2.46) clearly indicates the advantage of the system notation, since the expression for the determinant of an $N \times N$ matrix in the expanded form, even for relatively small values of N, is quite complex.

► Chapter 3

Vector Spaces

$\boxed{3.1}$ Introduction and Basic Concepts

In general, a mathematical space is a set of mathematical objects with an associated structure. This structure can be specified by a number of operations on the objects of the set. These operations must satisfy certain general rules, called the axioms of the mathematical space.

In order to specify the operations and axioms used to define a vector space, it is first important to introduce some general concepts. The set of values $(a^1, a^2, \ldots, a^N)$ of some N variables $(x^1, x^2, \ldots, x^N)$ is called a point. A set of all such points for all possible real values of the given N variables is called a real N-dimensional space. A vector in such a space is an ordered pair of points, such that it is specified which point is the first point (the origin of the vector) and which point is the last point (end of the vector). For example, if the point $P(a^1, a^2, \ldots, a^N)$ is the origin of the vector and the point $Q(b^1, b^2, \ldots, b^N)$ is the end of the vector, then the vector $\vec{PQ}$ is the vector of displacement from the point P to the point Q. The system of values

$$c^1 = b^1 - a^1, \quad c^2 = b^2 - a^2, \ldots, c^N = b^N - a^N, \qquad (3.1)$$

or in a more compact notation

$$c^i = b^i - a^i \quad (i = 1, 2, \ldots, N), \qquad (3.2)$$

is called the system of coordinates of the vector $\vec{PQ}$ in the N-dimensional space.

The vector $\vec{PQ}$ in the N-dimensional space is thus fully determined when we know all of the N coordinates of the point of origin P, i.e.,

$$a^i \quad (i = 1, 2, \ldots, N), \tag{3.3}$$

and all of the coordinates of the vector $\vec{PQ}$, given by (3.2). However, in many cases only the coordinates (3.2) are used to specify a vector, and it is assumed that they can be measured from an arbitrary point in the N-dimensional space as an origin. Such a vector is called a free vector. An arbitrary vector $\vec{PQ} = \vec{C}$ will therefore usually be identified with the system of coordinates

$$c^i \quad (i = 1, 2, \ldots, N). \tag{3.4}$$

If we adopt a common point of origin for all vectors, defined in the N-dimensional space, then the position of every point in that space is determined by its position vector with respect to the adopted common point of origin.

The point with coordinates $(0, 0, \ldots, 0)$ is called the origin of coordinates. It is customary to adopt it as the common point of origin for all vectors in an N-dimensional space, although it is not mandatory. The equality of two vectors $\vec{a}$ and $\vec{b}$ requires the equality of all of their components

$$a^i = b^i \quad (i = 1, 2, \ldots, N). \tag{3.5}$$

3.2 Definition of a Vector Space

A general N-dimensional space in which the operation of addition (subtraction) is defined by the equation

$$c^i = a^i \pm b^i \quad (i = 1, 2, \ldots, N), \tag{3.6}$$

and the operation of scalar multiplication is defined by the equation

$$a^i = \lambda b^i \quad (i = 1, 2, \ldots, N), \tag{3.7}$$

for an arbitrary scalar λ, is called *the vector space* in N dimensions. The two vector operations satisfy the following axioms:

1. The commutativity of addition:

$$a^i + b^i = b^i + a^i \quad (i = 1, 2, \ldots, N) \tag{3.8}$$

2. The associativity of addition:

$$(a^i + b^i) + c^i = a^i + (b^i + c^i) \quad (i = 1, 2, \ldots, N) \tag{3.9}$$

3. The existence of a null-vector 0^i such that

$$a^i + 0^i = 0^i + a^i \quad (i = 1, 2, \ldots, N) \tag{3.10}$$

4. The distribution laws for scalar multiplication:

$$\lambda(a^i + b^i) = \lambda a^i + \lambda b^i \quad (i = 1, 2, \ldots, N) \tag{3.11}$$

$$(\lambda + v)a^i = \lambda a^i + v a^i \quad (i = 1, 2, \ldots, N) \tag{3.12}$$

5. The associativity of scalar multiplication:

$$(\lambda v)a^i = \lambda(v a^i) \quad (i = 1, 2, \ldots, N) \tag{3.13}$$

6. The existence of a unit-scalar 1 which satisfies

$$1 a^i = a^i \quad (i = 1, 2, \ldots, N). \tag{3.14}$$

The foregoing definition of the N-dimensional vector space implies certain properties of objects in it. The colinear or parallel vectors in the N-dimensional vector space are the vectors $\vec{a}$ and $\vec{b}$ which are linearly dependent, i.e., such that

$$a^i = \lambda b^i \quad (i = 1, 2, \ldots, N). \tag{3.15}$$

A straight line in the vector space is defined by the equation

$$x^i = a^i + \lambda b^i \quad (i = 1, 2, \ldots, N), \tag{3.16}$$

where λ is the variable parameter. The vector $\vec{a}$ determines one point on the straight line, while the vector $\vec{b}$ determines the set of coefficients of the direction of the straight line. A plane in the vector space is defined by the equation

$$x^i = a^i + \lambda b^i + v c^i \quad (i = 1, 2, \ldots, N), \tag{3.17}$$

where λ and v are variable parameters. The vector $\vec{a}$ determines one point on the plane, while the vectors $\vec{b}$ and $\vec{c}$ are the direction vectors of the plane.

3.3   The Euclidean Metric Space

If the distance between two arbitrary points $P(a^1, a^2, \ldots, a^N)$ and $Q(b^1, b^2, \ldots, b^N)$ of a vector space is defined by the equation

$$s = \overline{PQ} = \left[(b^1 - a^1)^2 + (b^2 - a^2)^2 + \cdots + (b^N - a^N)^2 \right]^{1/2}, \quad (3.18)$$

such a vector space is called the *Euclidean metric space*. In the case of two infinitesimally close points $P(y^1, y^2, \ldots, y^N)$ and $Q(y^1 + dy^1, y^2 + dy^2, \ldots, y^N + dy^N)$, the distance ds is given by

$$ds^2 = (dy^1)^2 + (dy^2)^2 + \cdots + (dy^N)^2, \quad (3.19)$$

or

$$ds^2 = \delta_{kj} dy^k dy^j \quad (k, j = 1, 2, \ldots, N). \quad (3.20)$$

The expression (3.20) is called the square of the line element ds or the metric of the Euclidean metric space. The Euclidean metric space is a special case of the general vector space. Thus the operations of addition (subtraction) and scalar multiplication, satisfying the axioms (3.8)–(3.14), are defined in the Euclidean metric space. However, in a metric space there is another operation with vectors called the *scalar product*, defined as a composition of vectors a^j and b^j, i.e.,

$$\delta_{kj} a^k b^j = a^1 b^1 + a^2 b^2 + \cdots + a^N b^N. \quad (3.21)$$

The Euclidean metric space in N dimensions is usually denoted by E_N.

3.4   The Riemannian Spaces

In the previous section, we have concluded that in a Euclidean metric space E_N, the metric is defined by

$$ds^2 = \delta_{kj} dy^k dy^j \quad (k, j = 1, 2, \ldots, N), \quad (3.22)$$

where y^k are Descartes rectangular coordinates. If, instead of the Descartes coordinates y^k, we introduce N arbitrary generalized coordinates x^k, by means of the equations

$$y^k = y^k(x^1, x^2, \ldots, x^N) \quad (k = 1, 2, \ldots, N), \quad (3.23)$$

then, due to the relations

$$dy^k = \frac{\partial y^k}{\partial x^m} dx^m \quad (k, m = 1, 2, \ldots, N), \tag{3.24}$$

the metric can be written as follows:

$$ds^2 = \delta_{kj} \frac{\partial y^k}{\partial x^m} dx^m \frac{\partial y^j}{\partial x^n} dx^n = \delta_{kj} \frac{\partial y^k}{\partial x^m} \frac{\partial y^j}{\partial x^n} dx^m dx^n. \tag{3.25}$$

If we introduce a notation

$$g_{mn} = \delta_{kj} \frac{\partial y^k}{\partial x^m} \frac{\partial y^j}{\partial x^n}, \tag{3.26}$$

the metric can be written in the form

$$ds^2 = g_{mn} dx^m dx^n \quad (m, n = 1, 2, \ldots, N). \tag{3.27}$$

From (3.26) it is clear that g_{mn}, in general, is not equal to the δ-symbol and that (3.27) cannot be reduced to the sum of squares of differentials of N coordinates. Thus (3.27) is a general homogeneous quadratic form. Let us now, as an example, consider the usual three-dimensional Descartes system of coordinates $y^k = \{x, y, z\}$ and the system of spherical coordinates $x^k = \{r, \theta, \varphi\}$, defined by the equations

$$y^1 = x^1 \sin x^2 \cos x^3$$
$$y^2 = x^1 \sin x^2 \sin x^3 \tag{3.28}$$
$$y^3 = x^1 \cos x^2.$$

The components of the system g_{mn} in spherical coordinates are obtained using the definition (3.26) as follows:

$$g_{11} = \left(\frac{\partial y^1}{\partial x^1}\right)^2 + \left(\frac{\partial y^2}{\partial x^1}\right)^2 + \left(\frac{\partial y^3}{\partial x^1}\right)^2$$
$$= (\sin x^2 \cos x^3)^2 + (\sin x^2 \sin x^3)^2 + (\cos x^2)^2 = 1$$

$$g_{22} = \left(\frac{\partial y^1}{\partial x^2}\right)^2 + \left(\frac{\partial y^2}{\partial x^2}\right)^2 + \left(\frac{\partial y^3}{\partial x^2}\right)^2$$
$$= (x^1 \cos x^2 \cos x^3)^2 + (x^1 \cos x^2 \sin x^3)^2 + (-x^1 \sin x^2)^2 = (x^1)^2$$

$$g_{33} = \left(\frac{\partial y^1}{\partial x^3}\right)^2 + \left(\frac{\partial y^2}{\partial x^3}\right)^2 + \left(\frac{\partial y^3}{\partial x^3}\right)^2$$
$$= (-x^1 \sin x^2 \sin x^3)^2 + (x^1 \sin x^2 \cos x^3)^2 = (x^1 \sin x^2)^2. \tag{3.29}$$

Similar considerations show that

$$g_{mn} = 0 \quad \text{for } m \neq n. \tag{3.30}$$

Thus the metric (3.27) becomes

$$ds^2 = (dx^1)^2 + (x^1)^2(dx^2)^2 + (x^1 \sin x^2)^2(dx^3)^2, \tag{3.31}$$

or using the usual notation for the spherical coordinates

$$ds^2 = dr^2 + r^2 d\theta^2 + r^2 \sin^2 \theta d\varphi^2, \tag{3.32}$$

which is the well-known expression for the metric in spherical coordinates.

The expression (3.31) or (3.32) is a homogeneous quadratic form and it is not a sum of squares of differentials of the three coordinates.

In general a space of N dimensions, in which a metric is defined with respect to the Descartes system of rectangular coordinates (3.22), does not cease to be a Euclidean metric space when the metric is expressed with respect to some other generalized system of coordinates (3.27) and is no longer a sum of squares of differentials of N coordinates. The space remains the same Euclidean metric space and only the system of coordinates is changed.

Let us now consider a case when the Descartes coordinates y^k ($k = 1, 2, \ldots, N$) can be expressed in terms of a number M ($M < N$) of arbitrary variables x^α ($\alpha = 1, 2, \ldots, M$):

$$y^k = y^k(x^1, x^2, \ldots, x^M) \quad (k = 1, 2, \ldots, N). \tag{3.33}$$

In such a way we define an M-dimensional subspace, denoted by R_M, embedded in the original N-dimensional Euclidean metric space E_N. The metric of the Euclidean metric space has the form

$$ds^2 = \delta_{kj} dy^k dy^j \quad (k, j = 1, 2, \ldots, N), \tag{3.34}$$

and because of the relation

$$dy^k = \frac{\partial y^k}{\partial x^\alpha} dx^\alpha, \tag{3.35}$$

the metric of the subspace R_M can be written in the form

$$ds^2 = g_{\alpha\beta} dx^\alpha dx^\beta \quad (\alpha, \beta = 1, 2, \ldots, M), \tag{3.36}$$

where

$$g_{\alpha\beta} = \delta_{kj} \frac{\partial y^k}{\partial x^\alpha} \frac{\partial y^j}{\partial x^\beta}. \tag{3.37}$$

In (3.37) and below the Greek indices run from 1 to M while the Latin indices run from 1 to N.

Thus the square of the distance between two infinitesimally close points in the subspace R_M is defined by the homogeneous quadratic differential form (3.36). Let us now consider the M-dimensional subspace R_M independently of the original N-dimensional Euclidean metric space E_N, in which it is embedded. Now we would like to know whether it is possible to find M independent variables and use them as coordinates, such that the metric form (3.36) can be written as the sum of squares of differentials of these coordinates. If this is possible, the M-dimensional space R_M is in its own right a Euclidean metric space. If this is not possible the M-dimensional space R_M is called *the Riemannian space*.

Let us, for example, consider the points on a sphere of a unit radius in the three-dimensional Euclidean metric space E_3, defined by the Descartes rectangular coordinates $y^i = \{x, y, z\}$. The metric has the usual form

$$ds^2 = \delta_{kj} dy^k dy^j \quad (k, j = 1, 2, 3). \tag{3.38}$$

Let us now define a two-dimensional subspace R_2 of the Euclidean metric space E_3, in which the position of the points on the unit sphere is specified by the polar angles $x^\alpha = \{\theta, \varphi\}$. The relations (3.33) in this case have the form

$$
\begin{aligned}
y^1 &= \sin x^1 \cos x^2 \\
y^2 &= \sin x^1 \sin x^2 \\
y^3 &= \cos x^1.
\end{aligned}
\tag{3.39}
$$

The components of the system $g_{\alpha\beta}$ are obtained using the definition (3.37) as follows:

$$
\begin{aligned}
g_{11} &= \left(\frac{\partial y^1}{\partial x^1}\right)^2 + \left(\frac{\partial y^2}{\partial x^1}\right)^2 + \left(\frac{\partial y^3}{\partial x^1}\right)^2 \\
&= (\cos x^1 \cos x^2)^2 + (\cos x^1 \sin x^2)^2 + (-\sin x^1)^2 = 1 \\
g_{22} &= \left(\frac{\partial y^1}{\partial x^2}\right)^2 + \left(\frac{\partial y^2}{\partial x^2}\right)^2 + \left(\frac{\partial y^3}{\partial x^2}\right)^2 \\
&= (-\sin x^1 \sin x^2)^2 + (\sin x^1 \cos x^2)^2 = (\sin x^1)^2 \\
g_{12} &= g_{21} = 0.
\end{aligned}
\tag{3.40}
$$

Thus we obtain the metric of the subspace R_2 in the form

$$ds^2 = (dx^1)^2 + (\sin x^1)^2 (dx^2)^2, \tag{3.41}$$

or

$$ds^2 = d\theta^2 + \sin^2\theta d\varphi^2. \tag{3.42}$$

It turns out that it is *not possible* to find two real variables $\{u, v\}$ such that the metric (3.42) can be written in the form

$$ds^2 = du^2 + dv^2. \tag{3.43}$$

Thus the surface of the unit sphere, as a two-dimensional subspace R_2 of the original Euclidean metric space E_3, does not have the internal Euclidean metric. In other words, there are no Euclidean coordinates on the surface of the unit sphere.

It can, therefore, be concluded that in a Euclidean metric space there are some subspaces with a non-Euclidean metric. Such subspaces are called the *Riemannian spaces*. In principle, the metric geometry can be generalized by defining the metric of a space in advance by choosing the functions g_{mn} in an arbitrary way, with only requirements that the system g_{mn} be symmetric and doubly differentiable. The metric does not even have to be positively defined.

Thus a space, with a metric which is not positively defined but in which the components of the system g_{mn} are constants, is sometimes called a *pseudo-Euclidean space*. Analogously, a space which is not positively defined and in which the components of the system g_{mn} are arbitrary functions of coordinates is sometimes called a *pseudo-Riemannian space*.

▶ Chapter 4

Definitions of Tensors

4.1 Transformations of Variables

Tensors, as mathematical objects, were originally introduced for an immediate practical use in the theory of relativity. The main subject of the theory of relativity is the behavior of physical quantities and the laws of nature with respect to the transformations from one system of coordinates to another. It was therefore important to introduce a new class of mathematical objects that are defined by their transformation laws with respect to the transformations from one system of coordinates to another. Such mathematical objects are called *tensors*. The systems that we call tensors have linear and homogeneous transformation laws with respect to the transformations of coordinates. In order to define the main types of tensors, let us consider an arbitrary transformation of variables x^k into some new variables z^k, defined by

$$z^k = z^k(x^1, x^2, \ldots, x^N) \quad (k = 1, 2, \ldots, N). \tag{4.1}$$

From (4.1) we may write

$$dz^k = \frac{\partial z^k}{\partial x^m} dx^m \quad (k, m = 1, 2, \ldots, N). \tag{4.2}$$

The differentials of the coordinates dz^i are linear and homogeneous functions of the differentials of the old coordinates dx^m. The differentials of coordinates are by definition treated as the components of a special type of tensors, called *contravariant vectors*.

Let G be a zeroth-order system. The system of N quantities $\partial G/\partial x^i$ is transformed according to the transformation law

$$\frac{\partial G}{\partial z^k} = \frac{\partial x^m}{\partial z^k}\frac{\partial G}{\partial x^m} \quad (k, m = 1, 2, \ldots, N). \tag{4.3}$$

The components $\partial G/\partial z^k$ are linear and homogeneous functions of the components $\partial G/\partial x^m$. The components of the system $\partial G/\partial x^m$ are by definition treated as the components of a special type of tensors, called *covariant vectors*.

4.2 Contravariant Vectors

Generally speaking, any system of quantities, defined with respect to the systems of coordinates $\{x^k\}$ and $\{z^k\}$ by N quantities A^k and $\bar{A}^k$ respectively, which is transformed according to the transformation law

$$\bar{A}^k = \frac{\partial z^k}{\partial x^m}A^m \quad (k, m = 1, 2, \ldots, N) \tag{4.4}$$

is called a *contravariant vector*. The contravariant vectors are always denoted by one upper index.

4.3 Covariant Vectors

On the other hand, any system of quantities, defined with respect to the systems of coordinates $\{x^k\}$ and $\{z^k\}$ by N quantities B_k and $\bar{B}_k$, respectively, which is transformed according to the transformation law

$$\bar{B}_k = \frac{\partial x^m}{\partial z^k}B_m \quad (k, m = 1, 2, \ldots, N) \tag{4.5}$$

is called a *covariant vector*. The covariant vectors are always denoted by one lower index.

4.4 Invariants (Scalars)

Let us now form a zeroth-order system by composition of one contravariant and one covariant vector

$$F = A^m B_m \quad (m = 1, 2, \ldots, N), \tag{4.6}$$

with respect to a system of coordinates $\{x^k\}$. Then, with respect to a new system of coordinates $\{z^k\}$ this system has the value

$$\bar{F} = \bar{A}^k \bar{B}_k = \frac{\partial z^k}{\partial x^m}A^m \frac{\partial x^n}{\partial z^k}B_n = \frac{\partial z^k}{\partial x^m}\frac{\partial x^n}{\partial z^k}A^m B_n, \tag{4.7}$$

where the upper bar over the quantities $\bar{F}$, $\bar{A}^k$, and $\bar{B}_k$ denotes that they are defined with respect to the new coordinates $\{z^k\}$.

On the other hand, by definition of the δ-symbol, we have

$$\frac{\partial z^k}{\partial x^m} \frac{\partial x^n}{\partial z^k} = \delta^n_m \quad (k, m, n = 1, 2, \ldots, N), \tag{4.8}$$

and the result (4.7) becomes

$$\bar{F} = \delta^n_m A^m B_n = A^m B_m, \tag{4.9}$$

or

$$\bar{F} = \bar{A}^k \bar{B}_k = A^m B_m = F. \tag{4.10}$$

The result (4.10) shows that the quantity F has the same value in all systems of coordinates. Such a quantity is, therefore, called an *invariant* or a *scalar*. A composition of one contravariant and one covariant vector is therefore called the *scalar product* since it behaves as a scalar with respect to an arbitrary transformation of coordinates.

4.5 Contravariant Tensors

Let us consider a system of N^2 products of components of two contravariant vectors B^m and D^n, denoted by

$$A^{mn} = B^m D^n, \tag{4.11}$$

with respect to a system of coordinates $\{x^k\}$. In some other system of coordinates $\{z^k\}$ this system will have values in accordance with (4.4), i.e.,

$$\bar{A}^{jk} = \bar{B}^j \bar{D}^k = \frac{\partial z^j}{\partial x^m} \frac{\partial z^k}{\partial x^n} B^m D^n = \frac{\partial z^j}{\partial x^m} \frac{\partial z^k}{\partial x^n} A^{mn}. \tag{4.12}$$

In analogy with (4.12), any second-order system, defined with respect to the systems of coordinates $\{x^k\}$ and $\{z^k\}$ by N^2 quantities A^{mn} and $\bar{A}^{jk}$, respectively, that is transformed according to the transformation law

$$\bar{A}^{jk} = \frac{\partial z^j}{\partial x^m} \frac{\partial z^k}{\partial x^n} A^{mn} \tag{4.13}$$

is called a *second-order contravariant tensor*. The second-order contravariant tensors are always denoted by two upper indices.

A *third-order contravariant tensor* is a system of N^3 quantities, denoted by A^{mnp}, which is transformed according to the transformation law

$$\bar{A}^{ijk} = \frac{\partial z^i}{\partial x^m}\frac{\partial z^j}{\partial x^n}\frac{\partial z^k}{\partial x^p}A^{mnp}. \tag{4.14}$$

Analogously with the definitions (4.13) and (4.14), it is possible to define contravariant tensors of arbitrary order. By convention, any contravariant tensor is denoted by a number of upper indices only.

4.6   Covariant Tensors

A system defined with respect to the systems of coordinates $\{x^k\}$ and $\{z^k\}$ by N^2 quantities A_{mn} and $\bar{A}_{jk}$, respectively, that is transformed according to the transformation law

$$\bar{A}_{jk} = \frac{\partial x^m}{\partial z^j}\frac{\partial x^n}{\partial z^k}A_{mn} \tag{4.15}$$

is called a *second-order covariant tensor*. Second-order covariant tensors are always denoted by two lower indices.

A *third-order covariant tensor* is a system of N^3 quantities, denoted by A_{mnp}, which is transformed according to the transformation law

$$\bar{A}_{ijk} = \frac{\partial x^m}{\partial z^i}\frac{\partial x^n}{\partial z^j}\frac{\partial x^p}{\partial z^k}A_{mnp}. \tag{4.16}$$

Analogously with the definitions (4.15) and (4.16), it is possible to define covariant tensors of an arbitrary order. By convention, any covariant tensor is denoted by a number of lower indices only.

4.7   Mixed Tensors

Let us consider a system, defined with respect to the systems of coordinates $\{x^k\}$ and $\{z^k\}$ by N^4 quantities A^{mn}_{ps} and $\bar{A}^{ij}_{kl}$, respectively, which is transformed according to the transformation law

$$\bar{A}^{ij}_{kl} = \frac{\partial z^i}{\partial x^m}\frac{\partial z^j}{\partial x^n}\frac{\partial x^p}{\partial z^k}\frac{\partial x^s}{\partial z^l}A^{mn}_{ps}. \tag{4.17}$$

The system A^{mn}_{ps} is called a *fourth-order mixed tensor*, i.e., a second-order contravariant and second-order covariant tensor. The indices m, n are contravariant indices and indices p, s are covariant indices. In analogy with (4.17) it is possible to define mixed tensors with arbitrary numbers of contravariant and covariant indices. There are two more types of fourth-order mixed tensors

$$A^m_{nps}, \quad A^{mnp}_s. \tag{4.18}$$

Using the transformation law (4.17) it is easy to construct the transformation laws for arbitrary mixed tensors. The contravariant and covariant tensors can, of course, be treated as special cases of mixed tensors.

4.8 **Symmetry Properties of Tensors**

A second-order covariant tensor A_{jk} is called a *symmetric tensor* if its components satisfy the equality

$$A_{jk} = A_{kj} \quad (j, k = 1, 2, \ldots, N), \tag{4.19}$$

and the tensor A_{jk} is called an *antisymmetric tensor* if its components satisfy the equality

$$A_{jk} = -A_{kj} \quad (j, k = 1, 2, \ldots, N). \tag{4.20}$$

Analogously, a second-order contravariant tensor A^{jk} is called a *symmetric tensor* if its components satisfy the equality

$$A^{jk} = A^{kj} \quad (j, k = 1, 2, \ldots, N), \tag{4.21}$$

and the tensor A^{jk} is called an *antisymmetric tensor* if its components satisfy the equality

$$A^{jk} = -A^{kj} \quad (j, k = 1, 2, \ldots, N). \tag{4.22}$$

It is important to note that the symmetry properties of tensors are independent of the coordinate system in which the tensor components are defined. In other words, a tensor that is symmetric (antisymmetric) with respect to one coordinate system remains symmetric (antisymmetric) with respect to any other coordinate system.

In order to show that it is the case, let us assume that an arbitrary second-order contravariant tensor, defined with respect to the coordinate system $\{x^k\}$ by N^2 components A^{mn}, satisfies the equality

$$A^{mn} = A^{nm} \quad (m, n = 1, 2, \ldots, N), \tag{4.23}$$

which means that it is a symmetric tensor in the coordinate system $\{x^k\}$. In some other coordinate system $\{z^k\}$, this tensor is given by the components

$$\bar{A}^{jk} = \frac{\partial z^j}{\partial x^m} \frac{\partial z^k}{\partial x^n} A^{mn}. \tag{4.24}$$

However, by means of an interchange of the dummy indices $m \leftrightarrow n$, which is just a change of notation that does not affect the result of summation, and using (4.23) we obtain

$$\bar{A}^{jk} = \frac{\partial z^j}{\partial x^n} \frac{\partial z^k}{\partial x^m} A^{nm} = \frac{\partial z^k}{\partial x^m} \frac{\partial z^j}{\partial x^n} A^{mn} = \bar{A}^{kj}, \tag{4.25}$$

which shows that the tensor $\bar{A}^{jk}$ in the new coordinate system $\{z^k\}$ is indeed symmetric as well.

In general, if a tensor of an arbitrary order and type is symmetric (or antisymmetric) upon interchange of one pair of its indices (both lower or both upper indices) in one system of coordinates $\{x^k\}$, it remains symmetric (or antisymmetric) upon interchange of the corresponding pair of indices in any other system of coordinates $\{z^k\}$. In other words, the symmetry properties of tensors are independent of the coordinate system, in which the tensor components are defined.

4.9 Symmetric and Antisymmetric Parts of Tensors

Let us consider a second-order covariant tensor A_{mn}. This tensor can always be written as a sum of one symmetric and one antisymmetric tensor as follows:

$$A_{mn} = \frac{1}{2}(A_{mn} + A_{nm}) + \frac{1}{2}(A_{mn} - A_{nm}), \tag{4.26}$$

or

$$A_{mn} = A_{(mn)} + A_{[mn]}. \tag{4.27}$$

The symmetric tensor defined by the expression

$$A_{(mn)} = \frac{1}{2}(A_{mn} + A_{nm}) \tag{4.28}$$

is called the *symmetric part* of the tensor A_{mn}, while the antisymmetric tensor defined by the expression

$$A_{[mn]} = \frac{1}{2}(A_{mn} - A_{nm}) \tag{4.29}$$

is called the *antisymmetric part* of the tensor A_{mn}.

Analogously to (4.27), it is possible to use an arbitrary non-symmetric third-order covariant tensor A_{mnp} to create a totally symmetric and totally antisymmetric part as follows:

$$A_{(mnp)} = \frac{1}{3!}(A_{mnp} + A_{mpn} + A_{pmn} + A_{pnm} + A_{npm} + A_{nmp}) \qquad (4.30)$$

$$A_{[mnp]} = \frac{1}{3!}(A_{mnp} - A_{mpn} + A_{pmn} - A_{pnm} + A_{npm} - A_{nmp}) \qquad (4.31)$$

From (4.30) and (4.31) it is easily seen that any interchange of indices m, n, and p leaves $A_{(mnp)}$ unchanged, while it reverses the sign of $A_{[mnp]}$. The expressions (4.30) and (4.31) can be rewritten in a more compact form by introducing a special label for the permutations of the indices m, n, and p, as follows:

$$\pi_j(m, n, p) \quad (j = 0, 1, 2, 3, 4, 5). \qquad (4.32)$$

The components of the system (4.32) are the 3! permutations of the three indices m, n, and p, which can be listed as follows:

$$
\begin{aligned}
\pi_0(m, n, p) &= mnp \\
\pi_1(m, n, p) &= mpn \\
\pi_2(m, n, p) &= pmn \\
\pi_3(m, n, p) &= pnm \\
\pi_4(m, n, p) &= npm \\
\pi_5(m, n, p) &= nmp.
\end{aligned}
\qquad (4.33)
$$

Using (4.32) and (4.33), the expressions (4.30) and (4.31) can be rewritten in the following more compact form:

$$A_{(mnp)} = \frac{1}{3!} \sum_{j=0}^{3!-1} A_{\pi_j(mnp)} \qquad (4.34)$$

$$A_{[mnp]} = \frac{1}{3!} \sum_{j=0}^{3!-1} (-1)^j A_{\pi_j(mnp)}, \qquad (4.35)$$

valid for the third-order covariant tensor A_{mnp}.

The expressions (4.34) and (4.35) are easily generalized to the case of the Nth-order nonsymmetric tensor, where the totally symmetric and totally antisymmetric parts are given by

$$A_{(i_1 i_2 \ldots i_N)} = \frac{1}{N!} \sum_{j=0}^{N!-1} A_{\pi_j(i_1 i_2 \ldots i_N)} \qquad (4.36)$$

$$A_{[i_1 i_2 \ldots i_N]} = \frac{1}{N!} \sum_{j=0}^{N!-1} (-1)^j A_{\pi_j(i_1 i_2 \ldots i_N)}. \qquad (4.37)$$

4.10 Tensor Character of Systems

Sometimes we do not readily know the transformation laws for all systems encountered in different expressions. In order to be able to determine whether a given system is a tensor or not, we need some criteria for determination of the tensor character of systems. These criteria can be based on the following statement:

If an expression $A^k B_k$ is invariant with respect to the coordinate transformations and we know that A^k transforms as a contravariant vector (or that B_k transforms as a covariant vector), then we know that the system B_k is a covariant vector (or that the system A^k is a contravariant vector).

In order to prove the foregoing statement, let us begin with the assumption that the expression $A^k B_k$ is an invariant, i.e.,

$$\bar{A}^j \bar{B}_j - A^k B_k = 0. \qquad (4.38)$$

If we then know that A^j is a contravariant vector, we know that it transforms as

$$\bar{A}^j = \frac{\partial z^j}{\partial x^k} A^k. \qquad (4.39)$$

Substituting (4.39) into (4.38) we obtain

$$\frac{\partial z^j}{\partial x^k} A^k \bar{B}_j - A^k B_k = A^k \left(\frac{\partial z^j}{\partial x^k} \bar{B}_j - B_k \right) = 0. \qquad (4.40)$$

As the equality (4.40) is valid for an arbitrary contravariant vector A^k, we have

$$\frac{\partial z^j}{\partial x^k} \bar{B}_j - B_k = 0, \qquad (4.41)$$

or

$$\bar{B}_j = \frac{\partial x^k}{\partial z^j} B_k. \tag{4.42}$$

The expression (4.42) is the transformation law for a covariant vector B_k, and it shows that B_k is indeed a covariant vector. In the same way it can be shown that, if the expression $A_{mn}B^m D^n$ is invariant with respect to the coordinate transformations and we know that B^m and D^n are two different contravariant vectors, then A_{mn} is the second-order covariant tensor. This is valid even if, as a special case, the expression $A_{mn}B^m B^n$ is an invariant for an arbitrary contravariant vector B^k, provided that it is known that A_{mn} also satisfies the condition of symmetry

$$A_{mn} = A_{nm} \quad (m, n = 1, 2, \ldots, N). \tag{4.43}$$

These conditions can be generalized to tensors of an arbitrary order and used as the criteria for determination of the tensor character of systems.

As an example of these rules, let us consider a mixed second-order system δ_j^i. Let us take the numbers δ_j^i as the coordinates of a second-order mixed tensor with respect to an arbitrary coordinate system $\{x^k\}$, which is always possible. The question is whether they will keep their values, i.e., whether they will remain the coordinates of a δ-symbol, after the transformation to some new coordinate system $\{z^k\}$. Thus we may write

$$\bar{\delta}_j^i = \delta_n^m \frac{\partial z^i}{\partial x^m} \frac{\partial x^n}{\partial z^j}, \tag{4.44}$$

or, since $\{z^k\}$ is a system of mutually independent coordinates, we have

$$\bar{\delta}_j^i = \frac{\partial z^i}{\partial x^m} \frac{\partial x^m}{\partial z^j} = \frac{\partial z^i}{\partial z^j} = \delta_j^i. \tag{4.45}$$

The result (4.45) shows that the δ-symbol is indeed a second-order mixed tensor, which has the same coordinates in all coordinate systems.

▶ Chapter 5

Relative Tensors

5.1   Introduction and Definitions

The tensors defined in the previous chapter are sometimes also called the absolute tensors, since there are systems of quantities which, upon the transformations of coordinates, transform according to the similar but somewhat more general laws. Such systems are called relative tensors or pseudotensors. Thus a fifth-order system, three times contravariant and twice covariant,

$$A^{i_1 i_2 i_3}_{j_1 j_2} \quad (i_1, i_2, i_3, j_1, j_2 = 1, 2, \ldots, N) \tag{5.1}$$

is defined as a *relative tensor* or pseudotensor of weight M, if it transforms according to the transformation law

$$\bar{A}^{i_1 i_2 i_3}_{j_1 j_2} = \left| \frac{\partial x^r}{\partial z^s} \right|^M A^{m_1 m_2 m_3}_{n_1 n_2} \frac{\partial z^{i_1}}{\partial x^{m_1}} \frac{\partial z^{i_2}}{\partial x^{m_2}} \frac{\partial z^{i_3}}{\partial x^{m_3}} \frac{\partial x^{n_1}}{\partial z^{j_1}} \frac{\partial x^{n_2}}{\partial z^{j_2}}. \tag{5.2}$$

In (5.2) we may introduce a notation

$$\Delta = \left| \frac{\partial x^r}{\partial z^s} \right| \tag{5.3}$$

for the Jacobian of transformation of the original coordinates $\{x^k\}$ into the new coordinates $\{z^k\}$. In analogy with the definition (5.2), it is possible to define the relative tensors of an arbitrary weight, order, and type. The concept of relative tensors includes the absolute tensors, defined in the

previous chapter, as a special case. The absolute tensors can be treated as
relative tensors of weight zero.

In particular, the relative tensors of weight $M = +1$ are called the
tensor densities, while the relative tensors of weight $M = -1$ are called
the tensor capacities.

5.2 Unit Antisymmetric Tensors

As an important example of relative tensors, we consider the e-symbol in
three dimensions with three upper indices, i.e.,

$$e^{ijk} \quad (i,j,k = 1,2,3). \tag{5.4}$$

If we assume that e^{ijk} is a relative tensor of an unknown weight M, then it
transforms according to the transformation law

$$\bar{e}^{rst} = \left| \frac{\partial x^n}{\partial z^m} \right|^M e^{ijk} \frac{\partial z^r}{\partial x^i} \frac{\partial z^s}{\partial x^j} \frac{\partial z^t}{\partial x^k}. \tag{5.5}$$

On the other hand, by definition of the Jacobian of the transformation Δ,
we have

$$\Delta = \left| \frac{\partial x^n}{\partial z^m} \right| = \begin{vmatrix} \frac{\partial x^1}{\partial z^1} & \frac{\partial x^1}{\partial z^2} & \frac{\partial x^1}{\partial z^3} \\ \frac{\partial x^2}{\partial z^1} & \frac{\partial x^2}{\partial z^2} & \frac{\partial x^2}{\partial z^3} \\ \frac{\partial x^3}{\partial z^1} & \frac{\partial x^3}{\partial z^2} & \frac{\partial x^3}{\partial z^3} \end{vmatrix} = e_{ijk} \frac{\partial x^i}{\partial z^1} \frac{\partial x^j}{\partial z^2} \frac{\partial x^k}{\partial z^3}. \tag{5.6}$$

The Jacobian of the inverse transformation is given by

$$\Delta^{-1} = \left| \frac{\partial z^m}{\partial x^n} \right| = e^{ijk} \frac{\partial z^1}{\partial x^i} \frac{\partial z^2}{\partial x^j} \frac{\partial z^3}{\partial x^k}. \tag{5.7}$$

Let us now, for the moment, consider the following system:

$$A^{rst} = e^{ijk} a_i^r a_j^s a_k^t \tag{5.8}$$

for an arbitrary mixed system a_n^m. In the expanded form this system
looks like

$$A^{rst} = a_1^r a_2^s a_3^t - a_1^r a_3^s a_2^t + a_3^r a_1^s a_2^t - a_3^r a_2^s a_1^t + a_2^r a_3^s a_1^t - a_2^r a_1^s a_3^t. \tag{5.9}$$

From (5.9), we see that the system A^{rst} is a fully antisymmetric system with
respect to its three indices, in the space of three dimensions. However,
an arbitrary third-order antisymmetric system in three dimensions has
only one independent component, i.e., A^{123}, whereas the other five

nonzero components are determined by the conditions of full antisymmetry. Thus we may write

$$A^{rst} = A^{123} e^{rst}. \tag{5.10}$$

In other words, any fully antisymmetric third-order system in three dimensions is proportional to the corresponding e-symbol. Using (5.8) and (5.10) we obtain

$$e^{ijk} a_i^r a_j^s a_k^t = e^{rst} e^{ijk} a_i^1 a_j^2 a_k^3. \tag{5.11}$$

Using (5.11) we may write

$$e^{ijk} \frac{\partial z^r}{\partial x^i} \frac{\partial z^s}{\partial x^j} \frac{\partial z^t}{\partial x^k} = e^{rst} e^{ijk} \frac{\partial z^1}{\partial x^i} \frac{\partial z^2}{\partial x^j} \frac{\partial z^3}{\partial x^k}, \tag{5.12}$$

or using (5.7)

$$e^{ijk} \frac{\partial z^r}{\partial x^i} \frac{\partial z^s}{\partial x^j} \frac{\partial z^t}{\partial x^k} = e^{rst} \Delta^{-1}. \tag{5.13}$$

Substituting (5.13) into (5.5) we obtain

$$\bar{e}^{rst} = \Delta^M e^{rst} \Delta^{-1}. \tag{5.14}$$

From (5.14) we see that if we choose $M = +1$, the components of the system e^{rst} remain unchanged upon the coordinate transformations. Thus the system e^{ijk} is a third-order relative contravariant tensor with the weight $M = +1$ (tensor density). The transformation law of the unit antisymmetric system e^{ijk} is therefore given by

$$\bar{e}^{rst} = \left| \frac{\partial x^n}{\partial z^m} \right| e^{ijk} \frac{\partial z^r}{\partial x^i} \frac{\partial z^s}{\partial x^j} \frac{\partial z^t}{\partial x^k} = \Delta e^{ijk} \frac{\partial z^r}{\partial x^i} \frac{\partial z^s}{\partial x^j} \frac{\partial z^t}{\partial x^k}. \tag{5.15}$$

In a similar way, we can consider the e-symbol with three lower indices in three dimensions

$$e_{ijk} \quad (i, j, k = 1, 2, 3). \tag{5.16}$$

Again, if we assume that e_{ijk} is a relative tensor of an unknown weight M, then it transforms according to the transformation law

$$\bar{e}_{rst} = \left| \frac{\partial x^n}{\partial z^m} \right|^M e_{ijk} \frac{\partial x^i}{\partial z^r} \frac{\partial x^j}{\partial z^s} \frac{\partial x^k}{\partial z^t}. \tag{5.17}$$

Using an analog of Equation (5.11) in the form

$$e_{ijk} a_r^i a_s^j a_t^k = e_{rst} e_{ijk} a_1^i a_2^j a_3^k, \tag{5.18}$$

we obtain

$$e_{ijk}\frac{\partial x^i}{\partial z^r}\frac{\partial x^j}{\partial z^s}\frac{\partial x^k}{\partial z^t} = e_{rst}e_{ijk}\frac{\partial x^i}{\partial z^1}\frac{\partial x^j}{\partial z^2}\frac{\partial x^k}{\partial z^3}, \qquad (5.19)$$

or

$$e_{ijk}\frac{\partial x^i}{\partial z^r}\frac{\partial x^j}{\partial z^s}\frac{\partial x^k}{\partial z^t} = e_{rst}\Delta. \qquad (5.20)$$

Substituting (5.20) into (5.17) we obtain

$$\bar{e}_{rst} = \Delta^M e_{rst}\Delta. \qquad (5.21)$$

From (5.21) we see that if we choose $M = -1$, the components of the system e_{rst} remain unchanged upon the coordinate transformations. Thus the system e_{ijk} is a third-order relative contravariant tensor with the weight $M = -1$ (tensor capacity). The transformation law of the unit antisymmetric system e_{ijk} is therefore given by

$$\bar{e}_{rst} = \left|\frac{\partial x^n}{\partial z^m}\right|^{-1} e_{ijk}\frac{\partial x^i}{\partial z^r}\frac{\partial x^j}{\partial z^s}\frac{\partial x^k}{\partial z^t} = \Delta^{-1}e_{ijk}\frac{\partial x^i}{\partial z^r}\frac{\partial x^j}{\partial z^s}\frac{\partial x^k}{\partial z^t}. \qquad (5.22)$$

In general, the transformation law of an arbitrary tensor is defined by

1. Order—the number of indices

2. Type—the position and order of indices

3. Weight—the exponent of the Jacobian of transformation.

5.3   **Vector Product in Three Dimensions**

Let us consider two vectors of the same type, e.g., two contravariant vectors denoted by A^i and B^i, in a three-dimensional space ($i = 1, 2, 3$). Using the relative tensor e_{ijk}, we can define a first-order system

$$C_i = e_{ijk}A^j B^k \quad (i, j, k = 1, 2, 3), \qquad (5.23)$$

which is a first-order covariant relative vector with the weight $M = -1$, since it is defined as a composition of one relative tensor of weight $M = -1$ with two absolute vectors. The expression (5.23) can be expanded into three equations:

$$C_1 = A^2 B^3 - A^3 B^2$$
$$C_2 = A^3 B^1 - A^1 B^3 \qquad (5.24)$$
$$C_3 = A^1 B^2 - A^2 B^1.$$

The transformation law for the vector C_i is given by

$$\bar{C}_j = \Delta^{-1} C_p \frac{\partial x^p}{\partial z^j}, \qquad (5.25)$$

where the Jacobian of the transformation is given by

$$\Delta = \left| \frac{\partial x^n}{\partial z^m} \right|. \qquad (5.26)$$

In a similar way we may define a relative contravariant vector C^i of the weight $M = +1$, starting with two absolute covariant vectors A_j and B_j as well as the relative tensor e^{ijk}, i.e.,

$$C^i = e^{ijk} A_j B_k \quad (i, j, k = 1, 2, 3). \qquad (5.27)$$

The expression (5.27) can be expanded into three equations:

$$C^1 = A_2 B_3 - A_3 B_2$$
$$C^2 = A_3 B_1 - A_1 B_3 \qquad (5.28)$$
$$C^3 = A_1 B_2 - A_2 B_1.$$

The transformation law for the vector C^i is given by

$$\bar{C}^j = \Delta C^p \frac{\partial z^j}{\partial x^p}. \qquad (5.29)$$

The product (5.23) or (5.27) is called the *vector product* of two contravariant or two covariant vectors in three dimensions, respectively. This definition includes the usual definition of the vector product of two vectors (assumed to be the position vectors) given by their Descartes rectangular coordinates:

$$\vec{C} = \vec{A} \times \vec{B} = \begin{vmatrix} \vec{1} & \vec{2} & \vec{3} \\ A_1 & A_2 & A_3 \\ B_1 & B_2 & B_3 \end{vmatrix}. \qquad (5.30)$$

In the definition (5.30), $\vec{1}$, $\vec{2}$, and $\vec{3}$ are the unit vectors of the three mutually orthogonal axes of the Descartes coordinate system. It should be noted that, in the Descartes coordinate system, the contravariant and covariant coordinates of the vector product have the same numerical values, i.e.,

$$C^j = \delta^{jk} C_k. \tag{5.31}$$

5.4 Mixed Product in Three Dimensions

The mixed product of three contravariant vectors A^j, B^j, and C^j is formed by a composition of the vector product

$$D_j = e_{jkm} B^k C^m \quad (j, k, m = 1, 2, 3) \tag{5.32}$$

with the vector A^j. It has the form

$$V = A^j D_j = e_{jkm} A^j B^k C^m \quad (j, k, m = 1, 2, 3), \tag{5.33}$$

or, in the Descartes coordinate system,

$$V = \begin{vmatrix} A^1 & A^2 & A^3 \\ B^1 & B^2 & B^3 \\ C^1 & C^2 & C^3 \end{vmatrix}. \tag{5.34}$$

From the definition (5.33) it is evident that the zeroth-order system V is a relative invariant of the weight $M = -1$, or the scalar capacity, since it is composed of three absolute contravariant vectors and the tensor capacity e_{jkm}. The transformation law for this system is

$$\bar{V} = \Delta^{-1} V. \tag{5.35}$$

In a similar way, the mixed product of three covariant vectors A_j, B_j, and C_j is formed by composition of the vector product

$$D^j = e^{jkm} B_k C_m \quad (j, k, m = 1, 2, 3) \tag{5.36}$$

with the vector A_j, and it has the form

$$G = A_j D^j = e^{jkm} A_j B_k C_m \quad (j, k, m = 1, 2, 3), \tag{5.37}$$

or, in the Descartes orthogonal coordinates,

$$G = \begin{vmatrix} A_1 & A_2 & A_3 \\ B_1 & B_2 & B_3 \\ C_1 & C_2 & C_3 \end{vmatrix}. \tag{5.38}$$

From the definition (5.37) it is evident that the zeroth-order system G is a relative invariant of the weight $M = +1$, or the scalar density, since it is composed of three absolute covariant vectors and the tensor density e^{jkm}.

The transformation law for this system is

$$\bar{G} = \Delta G. \tag{5.39}$$

Unlike absolute scalars, the systems V and G, as relative scalars, in general change with respect to the transformations of coordinates. The behavior of these systems with respect to a special class of the orthogonal transformations of Descartes coordinates will be discussed in the next section.

5.5 Orthogonal Coordinate Transformations

In order to highlight some important properties of the vector products and mixed products in three dimensions, we will consider the orthogonal transformations of Descartes coordinates *rotation*, *translation*, and *inversion*.

5.5.1 Rotations of Descartes Coordinates

Let us observe two Descartes coordinate systems K and K' with a common z-axis, denoted by $\vec{3} = \vec{3}'$, perpendicular to the plane of the paper. The system K' is obtained as a result of a rotation in the positive sense of the system K about a common 3-axis for some angle θ, as shown in Figure 1.

The relation between the coordinates of a position vector of a given fixed point P, with respect to the coordinate systems K' and K, is given by

$$\begin{bmatrix} z^1 \\ z^2 \\ z^3 \end{bmatrix} = \begin{bmatrix} \cos\theta & \sin\theta & 0 \\ -\sin\theta & \cos\theta & 0 \\ 0 & 0 & 1 \end{bmatrix} \begin{bmatrix} x^1 \\ x^2 \\ x^3 \end{bmatrix}, \tag{5.40}$$

or

$$z^k = A_j^k x^j \quad (j, k = 1, 2, 3). \tag{5.41}$$

The transformation (5.41) is a linear transformation with the Jacobian

$$\Delta = \left| \frac{\partial z^k}{\partial x^j} \right| = \left| A_j^k \right| = 1. \tag{5.42}$$

Thus the Euclidean metric, given by the analog of Equation (3.18) in three dimensions, is invariant with respect to the rotations of the Descartes coordinates. This can easily be shown by using the distance between the point $P(x^1, x^2, x^3)$ from the origin $O(0, 0, 0)$, which in the coordinate

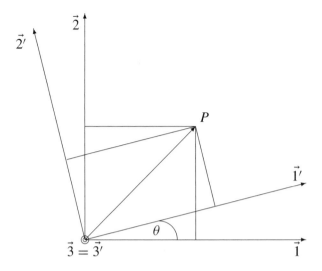

Figure 1. Rotations of Descartes coordinate systems

system K is given by

$$S = \left[(x^1)^2 + (x^2)^2 + (x^3)^2\right]^{1/2}. \tag{5.43}$$

In the coordinate system K' the distance between these points is given by

$$S' = \left[(z^1)^2 + (z^2)^2 + (z^3)^2\right]^{1/2}. \tag{5.44}$$

Substituting (5.40) into (5.44) we obtain

$$S' = \left[(x^1 \cos\theta + x^2 \sin\theta)^2 + (-x^1 \sin\theta + x^2 \cos\theta)^2 + (x^3)^2\right]^{1/2} \tag{5.45}$$

or

$$S' = S. \tag{5.46}$$

Thus the distance between the points in the Descartes coordinate system is invariant with respect to the rotation of coordinates about the 3-axis. The metric form of the space is also invariant with respect to the rotation of coordinates about the 3-axis, i.e., we have

$$ds'^2 = ds^2. \tag{5.47}$$

It is easily shown that the foregoing results are valid for arbitrary rotations of coordinates in the Descartes coordinate systems.

5.5.2   Translations of Descartes Coordinates

Let us consider two Descartes coordinate systems K and K' in three dimensions, where the coordinates in the system K' are denoted by $\{z^k\}$ and the coordinates in the system K are denoted by $\{x^k\}$. The translation of the coordinates is given by the equation

$$z^k = x^k + a^k \quad (k = 1, 2, 3), \tag{5.48}$$

where a^k is some constant translation vector. The Jacobian Δ of this transformation is given by

$$\Delta = \left| \frac{\partial z^k}{\partial x^j} \right| = \left| \delta_j^k \right| = 1. \tag{5.49}$$

Thus the metric form of the three-dimensional space in Descartes coordinates is invariant with respect to the translation of the coordinate systems, since we have

$$dz^k = dx^k \quad (k = 1, 2, 3). \tag{5.50}$$

From (5.50) we have

$$ds'^2 = \delta_{jk} dz^j dz^k = \delta_{jk} dx^j dx^k = ds^2 \quad (j, k = 1, 2, 3), \tag{5.51}$$

which proves the invariance of the metric form with respect to the coordinate translations.

5.5.3   Inversions of Descartes Coordinates

Let us again consider two Descartes coordinate systems K and K' in three dimensions, where the coordinates in the system K' are denoted by $\{z^k\}$ and the coordinates in the system K are denoted by $\{x^k\}$. The inversion of the coordinates is given by the equation

$$z^k = -x^k = -\delta_j^k x^j \quad (k = 1, 2, 3). \tag{5.52}$$

It can be shown that the inversion of Descartes coordinates cannot be achieved by means of any rotation of the original coordinate system K. The metric form is invariant with respect to the inversion, since we have

$$ds'^2 = \delta_{jk} dz^j dz^k = \delta_{jk} d(-x^j) d(-x^k) = \delta_{jk} dx^j dx^k = ds^2. \tag{5.53}$$

The Jacobian of this transformation is equal to

$$\Delta = \left| \frac{\partial z^k}{\partial x^j} \right| = \left| -\delta_j^k \right| = -1. \tag{5.54}$$

5.5.4   Axial Vectors and Pseudoscalars in Descartes Coordinates

Thus, as a conclusion of this section, we note that rotation, translation, and inversion constitute a group of orthogonal transformations which leave the metric form invariant. Furthermore, the transformation laws of relative vectors and scalars (e.g., vector products and mixed products in three dimensions), with respect to rotations and translations of coordinates, are the same as those for the absolute vectors and scalars. However, relative vectors and scalars (e.g., vector products and mixed products in three dimensions) do not transform like absolute vectors and scalars, with respect to the inversion of coordinates.

As we have concluded before, the transformation law of an absolute contravariant vector in three dimensions is given by

$$\bar{A}^j = \frac{\partial z^j}{\partial x^k} A^k \quad (j, k = 1, 2, 3), \tag{5.55}$$

with respect to the coordinate transformations from $\{x^k\}$ to $\{z^k\}$. An example of this type of vector is the polar vector of the position of a certain point in a three-dimensional space. However, the vector product of two absolute covariant vectors A_j and B_j is a relative contravariant vector C^j, which transforms according to the transformation law

$$\bar{C}^j = \Delta \frac{\partial z^j}{\partial x^k} C^k \quad (j, k = 1, 2, 3). \tag{5.56}$$

Substituting (5.42) or (5.49), into (5.56), we see that the vector product is transformed in the same way as the absolute vectors (5.55), with respect to rotations and translations. On the other hand, the vector product reverses sign with respect to the inversion of coordinates and does not transform as the polar vectors. This difference between the vector product, as the relative vector, and the polar vectors, being the absolute vectors, was noted in the three-dimensional vector algebra before the development of the tensor calculus. In the three-dimensional vector algebra, the vector product is called the *axial vector*, as opposed to the position vector which is called the *polar vector*. From the tensor point of view this difference is easily understood, since it relates to the definition of the position vector as an absolute vector and the vector product as a relative vector.

The mixed product in three dimensions, as a relative scalar, is invariant with respect to rotation and translation but it reverses sign with respect to inversion. In the three-dimensional vector algebra these scalars are usually called *pseudoscalars*.

▶ Chapter 6

The Metric Tensor

6.1 | Introduction and Definitions

As we have seen in the previous chapters, in the Euclidean metric space it is possible to reduce the metric form to a sum of squares of the differentials of the coordinates, i.e., we may write

$$ds^2 = \delta_{jk}\, dy^j\, dy^k \quad (j,k = 1, 2, \ldots, N). \tag{6.1}$$

If, instead of the Descartes orthogonal coordinates $\{y^k\}$, we introduce the arbitrary generalized coordinates $\{x^k\}$ by means of the equations

$$y^k = y^k(x^1, x^2, \ldots, x^N) \quad (k = 1, 2, \ldots, N), \tag{6.2}$$

we may write

$$dy^k = \frac{\partial y^k}{\partial x^m} dx^m \quad (k, m = 1, 2, \ldots, N), \tag{6.3}$$

and the metric form can be written as follows:

$$ds^2 = g_{mn}\, dx^m\, dx^n \quad (m, n = 1, 2, \ldots, N), \tag{6.4}$$

where

$$g_{mn} = \delta_{jk} \frac{\partial y^j}{\partial x^m} \frac{\partial y^k}{\partial x^n} \tag{6.5}$$

is a symmetric second-order system. The square of the infinitesimal line element ds is by definition an invariant in all coordinate systems and dx^m is an absolute vector. Thus, using the criteria for the tensor character of systems and the fact that g_{mn} is a symmetric system, we conclude that g_{mn} is an absolute second-order covariant tensor. This tensor is called the *metric tensor*. The determinant of the matrix associated with the metric tensor is given by

$$g = |g_{mn}| = \left| \delta_{jk} \frac{\partial y^j}{\partial x^m} \frac{\partial y^k}{\partial x^n} \right| = |\delta_{jk}| \left| \frac{\partial y^j}{\partial x^m} \right| \left| \frac{\partial y^k}{\partial x^n} \right| = \left| \frac{\partial y^k}{\partial x^n} \right|^2 \qquad (6.6)$$

where the multiplication rule for determinants has been used.

Thus the determinant of the metric tensor is equal to the square of the Jacobian of the transformation from the given Descartes coordinates to the arbitrary generalized coordinates. Assuming that the Jacobian of the transformation from the given Descartes coordinates to the arbitrary generalized coordinates is a nonzero real number, the determinant of the metric tensor is always a positive quantity, i.e., we have $g > 0$.

As an example, using the results (3.29), the matrix form of the metric tensor in the system of spherical coordinates $x^k = \{r, \theta, \varphi\}$ is given by

$$[g_{mn}] = \begin{bmatrix} 1 & 0 & 0 \\ 0 & (x^1)^2 & 0 \\ 0 & 0 & (x^1 \sin x^2)^2 \end{bmatrix}, \qquad (6.7)$$

and the determinant of the metric tensor is given by

$$g = \begin{vmatrix} 1 & 0 & 0 \\ 0 & (x^1)^2 & 0 \\ 0 & 0 & (x^1 \sin x^2)^2 \end{vmatrix} = \left[(x^1)^2 \sin x^2 \right]^2 = (r^2 \sin \theta)^2. \qquad (6.8)$$

As a system, the determinant g is a relative scalar invariant of the weight $M = 2$, which is easily shown as follows:

$$\bar{g} = \left| \frac{\partial y^k}{\partial z^j} \right|^2 = \left| \frac{\partial y^k}{\partial x^m} \frac{\partial x^m}{\partial z^j} \right|^2 = \left| \frac{\partial y^k}{\partial x^m} \right| \left| \frac{\partial x^m}{\partial z^j} \right|^2 = g \left| \frac{\partial x^m}{\partial z^j} \right|^2 \qquad (6.9)$$

or

$$\bar{g} = g\Delta^2, \qquad (6.10)$$

which proves that the determinant g is indeed a relative scalar invariant of the weight $M = 2$. From (6.10) we see that $\sqrt{g}$ is a relative scalar invariant of the weight $M = 1$, i.e.,

$$\sqrt{\bar{g}} = \sqrt{g}\Delta. \qquad (6.11)$$

Let us use G^{mn} to denote the cofactor of the element g_{mn} in the determinant $|g_{mn}|$. Then according to the determinant calculation rules, we have

$$g = g_{mn}G^{mn}. \tag{6.12}$$

The adjunct matrix to the matrix $[g_{mn}]$, which is denoted by adj $[g_{mn}]$, is by definition the transposed matrix of the cofactors, i.e., $[G^{mn}]^T = [G^{nm}]$. The matrix, inverse to the matrix $[g_{mn}]$, is therefore given by

$$\text{inv} \, [g_{mn}] = \frac{\text{adj} \, [g_{mn}]}{g} = \frac{[G^{mn}]^T}{g}. \tag{6.13}$$

By definition, a product of a matrix with its own inverse is equal to a unit matrix. Thus we may write

$$[g_{mn}]\frac{[G^{mn}]^T}{g} = 1, \tag{6.14}$$

or, using the system notation,

$$g_{mn}\frac{(G^{kn})^T}{g} = g_{mn}\frac{G^{nk}}{g} = \delta_m^k. \tag{6.15}$$

Introducing the notation

$$g^{nk} = \frac{G^{nk}}{g}, \tag{6.16}$$

the expression (6.15) becomes

$$g_{mn}g^{nk} = \delta_m^k. \tag{6.17}$$

From (6.17), we see that the system g^{nk} is a second-order contra-variant tensor, which is usually called the *contravariant metric tensor*. Analogously to the covariant metric tensor, the contravariant metric tensor is also a symmetric tensor, i.e., we have

$$g^{mn} = g^{nm}. \tag{6.18}$$

Using (6.17) and the multiplication rules for the determinants we find

$$|g_{mn}| \, \left|g^{nk}\right| = \left|\delta_m^k\right| = 1, \tag{6.19}$$

or

$$\left|g^{nk}\right| = \frac{1}{|g_{mn}|} = \frac{1}{g}. \tag{6.20}$$

The determinant of the contravariant metric tensor is a relative scalar invariant of the weight $M = -2$, which is easily shown using (6.10),

$$\frac{1}{\bar{g}} = \frac{1}{g}\Delta^{-2}, \tag{6.21}$$

or

$$\left|\bar{g}^{mn}\right| = \left|g^{mn}\right|\Delta^{-2}. \tag{6.22}$$

Since the antisymmetric unit system $e^{i_1 i_2 \cdots i_N}$ is a relative contravariant tensor of the weight $M = +1$, it is possible to form an absolute contravariant antisymmetric tensor $\varepsilon^{i_1 i_2 \cdots i_N}$, using $\sqrt{g}$, as follows:

$$\varepsilon^{i_1 i_2 \cdots i_N} = \frac{1}{\sqrt{g}} e^{i_1 i_2 \cdots i_N}. \tag{6.23}$$

On the other hand, since the antisymmetric unit system $e_{i_1 i_2 \cdots i_N}$ is a relative covariant tensor of the weight $M = -1$, it is possible to form an absolute unit covariant antisymmetric tensor $\varepsilon_{i_1 i_2 \cdots i_N}$, using $\sqrt{g}$, as follows:

$$\varepsilon_{i_1 i_2 \cdots i_N} = \sqrt{g}\, e_{i_1 i_2 \cdots i_N}. \tag{6.24}$$

The absolute tensors defined by (6.23) and (6.24) are called the *Ricci antisymmetric tensors* in the space of N dimensions.

6.2 Associated Vectors and Tensors

In a metric space, the contravariant and covariant tensors can be transformed to each other using the metric tensors g_{mn} and g^{mn}. In general the upper indices can be lowered and the lower indices can be made to be upper indices, using the metric tensors. For example, a covariant vector

$$A_m = g_{mn}A^n, \tag{6.25}$$

derived from a contravariant vector A^n using the metric tensor g_{mn}, is called the *associated vector* to the contravariant vector A^n. In the same way, a contravariant vector

$$A^m = g^{mn}A_n, \tag{6.26}$$

derived from a given covariant vector A_n using the metric tensor g^{mn}, is called the *associated vector* to the vector A_n.

For the associated vectors, the following rules are valid:

1. The association relation of two vectors is reciprocal. If a vector $A_m = g_{mn}A^n$ is associated to the vector A^n, then the associated vector to the vector A_m is the vector A^n. This can be shown as follows:

$$g^{nm}A_m = g^{nm}g_{mk}A^k = \delta_k^n A^k = A^n. \tag{6.27}$$

2. The absolute square of a contravariant vector A^m or a covariant vector A_m is the scalar (inner) product of a vector and its associated vector, i.e.,

$$g^{nm}A_m = g^{nm}g_{mk}A^k = \delta_k^n A^k = A^n. \tag{6.28}$$

3. From the preceding rule it is evident that the absolute squares of the associated vectors are equal to each other.

4. The scalar product of the vectors A^k and B_k is equal to the scalar product of the vectors A_m and B^m, and it is invariant with respect to an arbitrary coordinate transformation. This can be shown as follows:

$$g^{nm}A_m = g^{nm}g_{mk}A^k = \delta_k^n A^k = A^n. \tag{6.29}$$

These rules allow us to consider the vectors A_m and A^m as the covariant and contravariant coordinates of the same vector, which we may denote by $\vec{A}$. From (6.29) we see that, in the metric space, it is possible to define the scalar product of two vectors $\vec{A}$ and $\vec{B}$ regardless of their type, i.e.,

$$\vec{A} \cdot \vec{B} = A^k B_k = A_m B^m = g_{mn}A^m B^n. \tag{6.30}$$

The analogous rules apply to the tensors of an arbitrary order and type. By composition with the metric tensors g_{mn} and g^{mn}, the upper indices are lowered and the lower indices are turned to the upper indices, respectively. All tensors, created from each other by composition with one or more metric tensors, are called *associated tensors*. Thus, for an arbitrary second-order covariant tensor a_{mn} we can create three associated second-order tensors a^{mn}, a_n^m, and a_m^n, as follows:

$$
\begin{aligned}
a^{mn} &= g^{mk}g^{nj}a_{kj}\\
a_n^m &= g^{mk}a_{kn}\\
a_m^n &= g^{nk}a_{mk}.
\end{aligned}
\tag{6.31}
$$

It should be noted that the tensors a_n^m and a_m^n, associated to the tensor a_{mn}, as defined in (6.31), are in general not equal to each other. They are equal to each other only when the covariant tensor a_{mn} is symmetric, i.e., when $a_{mn} = a_{nm}$. If we apply (6.31) to the metric tensor itself, we first note that the definition of the associated contravariant metric tensor turns to identity:

$$g^{mn} = g^{mk} g^{nj} g_{kj} = g^{mk} \delta_k^n = g^{mn}. \tag{6.32}$$

Secondly, since the metric tensor is a symmetric tensor, there is a unique mixed metric tensor equal to the corresponding δ-symbol,

$$g_n^m = g^{mk} g_{kn} = \delta_n^m, \tag{6.33}$$

in agreement with the Equation (6.17).

6.3 | Arc Length of Curves: Unit Vectors

Let us consider a Riemannian metric space in N dimensions, described by a system of N generalized coordinates $\{x^i\}$. If these coordinates are functions of an arbitrary parameter t, then a curve in this Riemannian space may be specified by N parameter equations

$$x^k = x^k(t) \quad (k = 1, 2, \ldots, N). \tag{6.34}$$

The square of an infinitesimal *arc length* element of the curve between the points x^k and $x^k + dx^k$ is given by

$$ds^2 = g_{mn} \, dx^m \, dx^n \quad (m, n = 1, 2, \ldots, N). \tag{6.35}$$

The infinitesimal arc length element itself is given by

$$ds = \sqrt{g_{mn} \, dx^m \, dx^n} \quad (m, n = 1, 2, \ldots, N), \tag{6.36}$$

and its derivative with respect to the parameter is given by

$$\frac{ds}{dt} = \sqrt{g_{mn} \frac{dx^m}{dt} \frac{dx^n}{dt}} \quad (m, n = 1, 2, \ldots, N). \tag{6.37}$$

The *arc length of a curve* from some reference point, with a parameter value of t_0, to some arbitrary point, with a parameter value of t, is then equal to

$$s(t) = \int_{t_0}^{t} \sqrt{g_{mn} \frac{dx^m}{dt} \frac{dx^n}{dt}} \, dt. \tag{6.38}$$

From (6.35) we see that we may write

$$g_{mn}\frac{dx^m}{ds}\frac{dx^n}{ds} = g_{mn}\lambda^m\lambda^n = \lambda_m\lambda^m = \vec{\lambda}\cdot\vec{\lambda} = 1, \qquad (6.39)$$

where λ^m is a vector defined by the expression

$$\lambda^m = \frac{dx^m}{ds} \quad (m = 1, 2, \ldots, N). \qquad (6.40)$$

From (6.39) we see that vector λ^m has the absolute square equal to unity. Thus the absolute value of the vector λ^m, denoted by $|\lambda|$ is also equal to unity:

$$|\lambda| = \sqrt{\vec{\lambda}\cdot\vec{\lambda}} = 1. \qquad (6.41)$$

A vector with the absolute value equal to unity is called a *unit vector*. The vector λ^m is a unit contravariant vector. As x^m are the coordinates of a point on a given curve and s is the length of the arc of that curve, the vector λ^m is the *tangent unit vector* to this curve.

6.4 Angles between Vectors

Let us now consider two polar unit vectors λ^m and μ^m with a common origin and with the end points A and B, as shown in Figure 2.

From Figure 2 we see that

$$\epsilon^m = \mu^m - \lambda^m \quad (m = 1, 2, \ldots, N). \qquad (6.42)$$

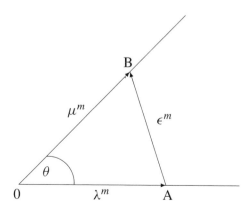

Figure 2. The angle θ between unit vectors λ^m and μ^m

Using the cosine theorem, the absolute square of the vector ϵ^m can be written in the form

$$|\epsilon|^2 = |\mu|^2 + |\lambda|^2 - 2\,|\mu|\,|\lambda| \cos\theta = 2 - 2\cos\theta = 2(1 - \cos\theta). \tag{6.43}$$

On the other hand, by definition, the absolute square of the vector ϵ^m has the form

$$|\epsilon|^2 = g_{mn}\epsilon^m\epsilon^n = g_{mn}(\mu^m - \lambda^m)(\mu^n - \lambda^n), \tag{6.44}$$

or

$$|\epsilon|^2 = g_{mn}\mu^m\mu^n + g_{mn}\lambda^m\lambda^n - 2g_{mn}\lambda^m\mu^n. \tag{6.45}$$

Since λ and μ are unit vectors, we have

$$g_{mn}\mu^m\mu^n = g_{mn}\lambda^m\lambda^n = 1, \tag{6.46}$$

and the result (6.45) becomes

$$|\epsilon|^2 = 2(1 - g_{mn}\lambda^m\mu^n). \tag{6.47}$$

Comparing the results (6.43) and (6.47) we find that the *cosine of an angle between two unit vectors* is given by the expression

$$\cos\theta = g_{mn}\lambda^m\mu^n. \tag{6.48}$$

If we have two arbitrary contravariant vectors A^m and B^m, then they define two directions with unit vectors

$$\lambda^m = \frac{A^m}{A}, \quad \mu^m = \frac{B^n}{B}. \tag{6.49}$$

The angle between these two directions is, according to (6.48), given by

$$\cos\theta = g_{mn}\frac{A^m B^n}{AB}. \tag{6.50}$$

From (6.50) we also see that the orthogonality condition for two contravariant vectors A^m and B^n has the form

$$g_{mn}A^m B^n = 0 \quad (m, n = 1, 2, \ldots, N). \tag{6.51}$$

The angle formed by two curves at their intersection is an angle between their tangent unit vectors at the intersection point. For two curves given by the parameter equations $x^m = \psi^m(t)$ and $x^n = \varphi^n(t)$, the angle formed by these curves at their intersection point is given by

$$\cos \theta = \frac{g_{mn} \dfrac{d\psi^m}{dt} \dfrac{d\varphi^n}{dt}}{\sqrt{g_{mn} \dfrac{d\psi^m}{dt} \dfrac{d\psi^n}{dt}} \sqrt{g_{mn} \dfrac{d\varphi^m}{dt} \dfrac{d\varphi^n}{dt}}}, \tag{6.52}$$

where g_{mn}, $d\psi^m/dt$, and $d\varphi^m/dt$ are all calculated at the intersection point of the two curves. From (6.51) we see that the orthogonality condition for the two curves at their intersection point is given by

$$g_{mn}\psi^m\varphi^n = 0 \quad (m, n = 1, 2, \ldots, N). \tag{6.53}$$

6.5 Schwarz Inequality

Let us consider two arbitrary contravariant vectors, denoted by A^m and B^m, and let us form a linear combination of these two vectors as follows:

$$A^m + \alpha B^m = C^m, \tag{6.54}$$

where α is an arbitrary real absolute scalar parameter. The linear combination C^m itself is also a contravariant vector. The absolute square of the vector C^m is given by the positive definite form

$$
\begin{aligned}
|C|^2 &= g_{mn}C^m C^n \\
|C|^2 &= g_{mn}(A^m + \alpha B^m)(A^n + \alpha B^n) \\
|C|^2 &= g_{mn}A^m A^n + 2\alpha g_{mn}A^m B^n + \alpha^2 g_{mn}B^m B^n \\
|C|^2 &= |A|^2 + 2\alpha g_{mn}A^m B^n + \alpha^2 |B|^2 .
\end{aligned}
\tag{6.55}
$$

For the quadratic form (6.55) to be nonnegative for all values of the parameter α, the following inequality must be satisfied:

$$(g_{mn}A^m B^n)^2 - |A|^2 |B|^2 \leq 0. \tag{6.56}$$

Using the notation $A = |A|$ and $B = |B|$, the inequality (6.56) gives

$$\left| g_{mn}A^m B^n \right| \leq AB. \tag{6.57}$$

The inequality (6.57) is known as the *Schwarz inequality* and it is of importance in a number of branches of mathematics and physics.

6.6 Orthogonal and Physical Vector Coordinates

In this section we consider an important special case when the three-dimensional Euclidean metric space is defined by some set of orthogonal curvilinear coordinates $\{x^k\}$ $(k = 1, 2, 3)$. For such a system of curvilinear coordinates, in every point of space the following conditions are satisfied:

$$g_{12} = g_{23} = g_{31} = 0. \tag{6.58}$$

Thus, we may write

$$g_{mn} = h_M^2 \delta_{mn} \quad (m, n, M = 1, 2, 3), \tag{6.59}$$

where $h_M = h_M(x^k)$ are some functions of coordinates $\{x^k\}$. In this case the metric of the space can be written in the form

$$ds^2 = (h_1 dx^1)^2 + (h_2 dx^2)^2 + (h_3 dx^3)^2, \tag{6.60}$$

and the matrix of the metric tensor is a diagonal 3×3 matrix

$$[g_{mn}] = \begin{bmatrix} (h_1)^2 & 0 & 0 \\ 0 & (h_2)^2 & 0 \\ 0 & 0 & (h_3)^2 \end{bmatrix}. \tag{6.61}$$

Let us now, as an example, consider the vector of generalized velocities in this coordinate system. The contravariant coordinates of the velocity vector are given by

$$v^m = \frac{dx^m}{dt}, \tag{6.62}$$

where dx^m are the coordinates of a contravariant polar vector in any system of coordinates and dt is an absolute scalar parameter. Using (6.61) we can calculate the covariant coordinates of the velocity vector as follows:

$$v_m = g_{mn} v^n = h_M^2 \delta_{mn} v^n. \tag{6.63}$$

As the generalized coordinates do not necessarily have the dimension of length (e.g., they can be the angular coordinates), the functions $h_M = h_M(x^k)$ are not necessarily dimensionless (e.g., they may have the dimension of length). From (6.63) it follows that the dimensions of the contravariant and covariant coordinates of the velocity vector are not the same, and they are not the same as the expected dimension of the velocity vector (i.e., length/time). The coordinates of the velocity vector that do have the dimension of velocity (i.e., length/time) are called the *physical coordinates* of the velocity vector. The physical coordinates can, in the special case under consideration, be obtained from the contravariant or covariant coordinates using the formulae

$$v_{(m)} = h_M v^m = \frac{1}{h_M} v_m. \tag{6.64}$$

As an illustration, let us consider the velocity vector in the spherical coordinates, where

$$x^1 = r, \quad x^2 = \theta, \quad x^3 = \varphi, \tag{6.65}$$

and

$$h_1 = 1, \quad h_2 = r, \quad h_3 = r \sin \theta. \tag{6.66}$$

The contravariant coordinates of the velocity vector in the spherical coordinate system are given by

$$v^1 = \frac{dr}{dt}, \quad v^2 = \frac{d\theta}{dt}, \quad v^3 = \frac{d\varphi}{dt}. \tag{6.67}$$

The associated covariant coordinates of the velocity vector in the spherical coordinate system are given by

$$v_1 = \frac{dr}{dt}, \quad v_2 = r^2 \frac{d\theta}{dt}, \quad v_3 = r^2 \sin^2 \theta \frac{d\varphi}{dt}. \tag{6.68}$$

From the results (6.67) and (6.68) we see that neither all the contravariant nor the covariant coordinates of the velocity vector have the dimension of velocity (i.e., length/time).

The physical coordinates of the velocity vector in the spherical coordinate system, i.e., the projections of the velocity vector to the directions of the curvilinear axes in a given space point, according to the definition (6.64) have the form

$$v_{(1)} = \frac{dr}{dt}, \quad v_{(2)} = r \frac{d\theta}{dt}, \quad v_{(3)} = r \sin \theta \frac{d\varphi}{dt}. \tag{6.69}$$

In a similar way, we can construct the physical coordinates of an arbitrary vector with respect to an arbitrary orthogonal system of coordinates in three dimensions. This approach can also be extended to an arbitrary N-dimensional space, but it is most commonly used in three or sometimes four dimensions.

Chapter 7

Tensors as Linear Operators

Second-order tensors can be described as linear operators acting on vectors in metric spaces. An operator in the N-dimensional metric space is defined by the way it acts on different vectors in the space. For example, in the expression

$$B^m = O^m_n A^n \quad (m, n = 1, 2, \ldots, N), \tag{7.1}$$

the tensor O^m_n represents a *linear operator*, which in a unique way relates a vector B^m to the original vector A^m. This operator is called a linear operator since it satisfies the conditions of linearity and homogeneity, i.e.,

$$\begin{aligned} O^m_n (A^n + B^n) &= O^m_n A^n + O^m_n B^n \\ O^m_n (\beta A^n) &= \beta O^m_n A^n, \end{aligned} \tag{7.2}$$

where A^n and B^n are arbitrary contravariant vectors and β is an arbitrary absolute scalar.

Eigenvectors of an operator O^m_n are defined as those vectors S^n that retain the direction and only change the absolute value as a result of the action of the operator O^m_n. In other words, for the eigenvectors S^n, we have

$$O^m_n S^n = \lambda S^m \quad (m, n = 1, 2, \ldots, N), \tag{7.3}$$

or

$$(O^m_n - \lambda \delta^m_n) S^n = 0 \quad (m, n = 1, 2, \ldots, N). \tag{7.4}$$

The system of homogeneous linear Equations (7.4) has a non-trivial solution for S^n only if its determinant is equal to zero, i.e.,

$$\det(O_n^m - \lambda \delta_n^m) = 0. \tag{7.5}$$

The N solutions for the parameter λ of the algebraic Equation (7.5) are called the *eigenvalues* of the operator O_n^m. The Equation (7.5) is sometimes called the *secular equation*.

As an example of the concepts we have just defined, let us consider a tensor O_n^m, defined in a three-dimensional metric space by the matrix

$$[O_n^m] = \begin{bmatrix} 1 & 0 & 5 \\ 0 & -2 & 0 \\ 5 & 0 & 1 \end{bmatrix}. \tag{7.6}$$

Substituting (7.6) into (7.5) we obtain the equation for the parameter λ in the form

$$\begin{vmatrix} 1-\lambda & 0 & 5 \\ 0 & -2-\lambda & 0 \\ 5 & 0 & 1-\lambda \end{vmatrix} = 0. \tag{7.7}$$

By expanding the determinant in (7.7) we obtain

$$\begin{aligned} (1-\lambda)&[(-2-\lambda)(1-\lambda)] + 5[-5(-2-\lambda)] \\ &= (1-\lambda)[(\lambda+2)(\lambda-1)] + 25(\lambda+2)] \\ &= (\lambda+2)[25 - (\lambda-1)^2] = 0. \end{aligned} \tag{7.8}$$

The solutions of the Equation (7.8) for the parameter λ are the following:

$$\begin{aligned} \lambda + 2 = 0 &\Rightarrow \lambda = -2 \\ 25 - (\lambda - 1)^2 = 0 &\Rightarrow \lambda = 1 \pm 5. \end{aligned} \tag{7.9}$$

Thus the three eigenvalues of the operator O_n^m are equal to

$$\lambda_S = -2, \quad \lambda_P = -4, \quad \lambda_Q = +6. \tag{7.10}$$

The corresponding eigenvectors S^n, P^n, and Q^n are obtained from the matrix equations

$$\begin{bmatrix} 1 & 0 & 5 \\ 0 & -2 & 0 \\ 5 & 0 & 1 \end{bmatrix} \begin{bmatrix} S^1 \\ S^2 \\ S^3 \end{bmatrix} = -2 \begin{bmatrix} S^1 \\ S^2 \\ S^3 \end{bmatrix}, \tag{7.11}$$

$$\begin{bmatrix} 1 & 0 & 5 \\ 0 & -2 & 0 \\ 5 & 0 & 1 \end{bmatrix} \begin{bmatrix} P^1 \\ P^2 \\ P^3 \end{bmatrix} = -4 \begin{bmatrix} P^1 \\ P^2 \\ P^3 \end{bmatrix}, \tag{7.12}$$

$$\begin{bmatrix} 1 & 0 & 5 \\ 0 & -2 & 0 \\ 5 & 0 & 1 \end{bmatrix} \begin{bmatrix} Q^1 \\ Q^2 \\ Q^3 \end{bmatrix} = +6 \begin{bmatrix} Q^1 \\ Q^2 \\ Q^3 \end{bmatrix}. \tag{7.13}$$

By solving the matrix Equations (7.11)–(7.13), we obtain the normalized (unit) eigenvectors of the operator O_n^m as follows:

$$[S^n] = \begin{bmatrix} S^1 \\ S^2 \\ S^3 \end{bmatrix} = \begin{bmatrix} 0 \\ 1 \\ 0 \end{bmatrix}, \tag{7.14}$$

$$[P^n] = \begin{bmatrix} P^1 \\ P^2 \\ P^3 \end{bmatrix} = \frac{1}{\sqrt{2}} \begin{bmatrix} 1 \\ 0 \\ -1 \end{bmatrix}, \tag{7.15}$$

$$[Q^n] = \begin{bmatrix} Q^1 \\ Q^2 \\ Q^3 \end{bmatrix} = \frac{1}{\sqrt{2}} \begin{bmatrix} 1 \\ 0 \\ 1 \end{bmatrix}. \tag{7.16}$$

It is easy to show by direct substitution that the eigenvectors (7.14)–(7.16) satisfy the above matrix Equations (7.11)–(7.13), respectively. It should also be noted that the eigenvectors S^n, P^n and Q^n, which correspond to different eigenvalues, are orthogonal to each other. In other words, it is easy to show by direct calculation that these eigenvectors satisfy the orthogonality conditions

$$[S^n]^T[P^n] = [S^n]^T[Q^n] = [P^n]^T[Q^n] = 0, \tag{7.17}$$

where the notation M^T is used for a transposed matrix of an arbitrary matrix M. Furthermore, the three eigenvectors are normalized so that their absolute values are equal to unity. In other words, they are unit vectors of the three independent directions. Thus, it can be shown by direct calculation that the eigenvectors satisfy the normalization conditions

$$[S^n]^T[S^n] = [P^n]^T[P^n] = [Q^n]^T[Q^n] = 1. \tag{7.18}$$

The set of normalized eigenvectors orthogonal to each other is called the set of *orthonormal eigenvectors*.

In general, the eigenvectors of an operator O_n^m, corresponding to the different eigenvalues, are mutually orthogonal if the tensor $O^{mn} = g^{mk} O_k^n$

is symmetric. This can be shown by using the definitions

$$O_n^m S^n = \lambda_S S^m$$
$$O_n^m P^n = \lambda_P P^m \tag{7.19}$$

or

$$O_n^m S^n P_m = g_{nk} O^{mk} S^n P_m = O^{mk} S_k P_m = \lambda_S S^m P_m$$
$$O_n^m P^n S_m = g_{nk} O^{mk} P^n S_m = O^{mk} P_k S_m = \lambda_P P^m S_m. \tag{7.20}$$

Using the symmetry of the tensor O^{mk}, i.e., the equality $O^{mk} = O^{km}$, we can interchange the dummy indices $m \leftrightarrow k$ to see that the left-hand sides of both of the Equations (7.20) are equal to each other. Furthermore, we note the equality $S^m P_m = P^m S_m$. With these properties, we can subtract the second equation from the first equation in (7.20) to obtain

$$0 = (\lambda_S - \lambda_P) S^m P_m. \tag{7.21}$$

From the result (7.21) we see that whenever the two eigenvalues are not equal to each other, i.e., whenever $\lambda_S \neq \lambda_P$, the two corresponding eigenvectors are indeed orthogonal to each other:

$$S^m P_m = g_{mn} S^m P^n = 0. \tag{7.22}$$

The eigenvectors of an operator are in general determined up to an arbitrary multiplication constant. However, in most cases it is convenient to have normalized eigenvectors of an operator, i.e., to use unit vectors as eigenvectors. Thus we require, as a convention, that the eigenvectors should satisfy the normalization conditions

$$S^m S_m = g_{mn} S^m S^n = 1. \tag{7.23}$$

This implies that the eigenvectors of an operator O_n^m in the three-dimensional metric space (e.g., S^m, P^m, and Q^m above) define three mutually orthogonal directions, provided that the corresponding eigenvalues are not equal to each other, i.e., $\lambda_S \neq \lambda_P \neq \lambda_Q$. These three directions are sometimes called the *main directions* of a second-order tensor O_n^m.

If the roots of the secular Equation (7.5) are not distinct, we have a *degeneracy*. In the three-dimensional case, if two roots are equal to each other, the second-order tensor has only one main direction and a plane perpendicular to it, where all directions are main directions of the tensor. If all three eigenvalues of a second-order tensor are equal to each other, such a tensor does not distinguish any particular directions in the three-dimensional metric space. In other words, all directions in the three-dimensional metric space are the main directions of such a tensor.

Part II

Tensor Analysis

▶ Chapter 8

Tensor Derivatives

8.1 Differentials of Tensors

In the curvilinear coordinates the differential of a covariant vector A^m, denoted by dA^m, is not a vector. It is easily shown using the transformation law of covariant vectors (4.5). If a covariant vector is defined with respect to the systems of coordinates $\{x^k\}$ and $\{z^k\}$ by N coordinates A_m and $\bar{A}_m$, respectively, then it transforms according to the transformation law

$$\bar{A}_m = \frac{\partial x^n}{\partial z^m} A_n \quad (m, n = 1, 2, \ldots, N). \tag{8.1}$$

Using (8.1) we obtain

$$d\bar{A}_m = \frac{\partial x^n}{\partial z^m} dA_n + d\left(\frac{\partial x^n}{\partial z^m}\right) A_n, \tag{8.2}$$

or

$$d\bar{A}_m = \frac{\partial x^n}{\partial z^m} dA_n + A_n \frac{\partial^2 x^n}{\partial z^m \partial z^k} dz^k. \tag{8.3}$$

The result (8.3) clearly shows that dA_n does not transform as a vector, except in a special case when

$$\frac{\partial^2 x^n}{\partial z^m \partial z^k} = 0, \tag{8.4}$$

i.e., in the case of a linear orthogonal transformation of Descartes coordinates into some new Descartes coordinates. As the differential of a covariant vector

$$dA_m = \frac{\partial A_m}{\partial x^n} dx^n \quad (m, n = 1, 2, \ldots, N) \tag{8.5}$$

is a system that is not a covariant vector, while the system dx^n is a covariant vector, the second-order system $\partial A_m/\partial x^n$ is not a tensor. The observation that the differential dA_m is not a vector is related to the definition of the differential, i.e.,

$$dA_m = A_m(x^k + dx^k) - A_m(x^k) \quad (k, m = 1, 2, \ldots, N). \tag{8.6}$$

By definition (8.6), dA_m is a difference between two covariant vectors with the origins in different, infinitesimally close points of the metric space. On the other hand, the multiplier in the transformation law (8.1) is of the form

$$\frac{\partial x^n}{\partial z^m} = f_m^n(x^k), \tag{8.7}$$

and it is in general a function of coordinates. The multiplier (8.7) is different in different points of the metric space, except in the special case of linear transformations. It is therefore evident that vectors, in general, transform differently in different points of the metric space, and consequently the differential (8.6) cannot transform as a covariant vector. In order to construct a differential of a covariant vector in curvilinear coordinates, which is a covariant vector itself, it is required that both vectors on the right-hand side of (8.6) be situated in the same point of the metric space. This can be achieved by moving one of the infinitesimally close vectors entering the right-hand side of the Equation (8.6) to the same point where the other vector is situated.

The operation of moving one of the infinitesimally close vectors to the point where the other vector is situated must be performed in such a way that in the Descartes coordinates the resulting difference is reduced to the ordinary differential dA_m. Since dA_m is a difference of the components of the two infinitesimally close vectors, the components of a vector to be moved to the point where the other vector is situated must remain unchanged. This can only be achieved by a parallel displacement of the vector between the two infinitesimally close points, using the Descartes coordinates.

Let us therefore consider an arbitrary contravariant vector with the coordinates A^m at a point of the metric space with coordinates x^m. The coordinates of this contravariant vector at the point of the metric space with coordinates $x^m + dx^m$ are denoted by $A^m + dA^m$. An infinitesimal translation of the vector A^m to the point of the metric space with coordinates $x^m + dx^m$ generates a translated vector denoted by $\tilde{A}^m$ at the point of the metric space with coordinates $x^m + dx^m$. Thus the differential of the contravariant vector A^m can be written in the form

$$dA^m = A^m(x^k + dx^k) - A^m(x^k)$$
$$dA^m = A^m(x^k + dx^k) - \tilde{A}^m(x^k + dx^k)$$
$$+ \tilde{A}^m(x^k + dx^k) - A^m(x^k) \tag{8.8}$$
$$dA^m = DA^m(x^k + dx^k) + \delta A^m.$$

In Equation (8.8) we have introduced a differential DA^m between the two vectors $A^m(x^k + dx^k)$ and $\tilde{A}^m(x^k + dx^k)$ situated at the same point $x^k + dx^k$ of the metric space, which itself behaves as a contravariant vector with respect to the coordinate transformations, as follows:

$$DA^m = A^m(x^k + dx^k) - \tilde{A}^m(x^k + dx^k). \tag{8.9}$$

Furthermore, in Equation (8.8) we have introduced an increment δA^m, due to the parallel translation of the vector A^m to the point of the metric space with coordinates $x^m + dx^m$, as follows:

$$\delta A^m = \tilde{A}^m(x^k + dx^k) - A^m(x^k). \tag{8.10}$$

As δA^m is a difference between the translated vector $\tilde{A}^m$ at the point $x^k + dx^k$ and the nontranslated vector A^m at the infinitesimally close point x^k, the system δA^m is not a vector. The difference δA^m vanishes in the Descartes coordinates and DA^m reduces to dA^m, as required.

In order to calculate the increment δA^m, we note that it is a function of the coordinates of the contravariant tensor A^m themselves. This functional dependence must be linear, since a sum of two vectors must transform according to the same transformation law as each of the vectors. Furthermore, it also has to be a linear function of the coordinate differentials. Thus, we may write

$$\delta A^m = -\Gamma^m_{np} A^n dx^p, \tag{8.11}$$

where Γ^m_{np} is a system of functions of coordinates, which are usually called the *Christoffel symbols of the second kind*. Composition with the metric tensor gives the *Christoffel symbols of the first kind*, as follows:

$$\Gamma_{m,np} = g_{mk}\Gamma^k_{np}. \tag{8.12}$$

The system Γ^m_{np} is dependent on the coordinate system and in Descartes coordinates all of its components vanish, i.e., $\Gamma^m_{np} = 0$. It is therefore clear that Γ^m_{np} is not a tensor, since a tensor that is equal to zero in one coordinate system must remain equal to zero in any other coordinate system. In a Riemannian space it is not possible to find a coordinate system to satisfy the condition

$$\Gamma^k_{np} = 0 \quad (k, n, p = 1, 2, \ldots, N) \tag{8.13}$$

in the entire metric space.

8.1.1 Differentials of Contravariant Vectors

Substituting the result (8.11) into the definition (8.8), we obtain the result for the differential of a contravariant vector A^m, as follows:

$$DA^m = dA^m - \delta A^m = \frac{\partial A^m}{\partial x^p} dx^p + \Gamma^m_{np} A^n dx^p, \tag{8.14}$$

or

$$DA^m = \left(\frac{\partial A^m}{\partial x^p} + \Gamma^m_{np} A^n \right) dx^p = \frac{DA^m}{Dx^p} dx^p. \tag{8.15}$$

8.1.2 Differentials of Covariant Vectors

In order to derive an expression analogous to the result (8.15) for covariant vectors, let us consider an absolute covariant vector A_m and an absolute contravariant vector B^m. The composition of these two vectors gives an absolute scalar $A_m B^m$. As the scalars are invariant with respect to the parallel translation, we can write

$$\delta(A_m B^m) = B^m \delta A_m + A_m \delta B^m = 0, \tag{8.16}$$

or

$$B^m \delta A_m = -A_m \delta B^m = +A_n \Gamma^n_{mp} B^m dx^p, \tag{8.17}$$

where we have made an interchange of the dummy indices $m \leftrightarrow n$ on the right-hand side of Equation (8.17), such that it can be rewritten as

$$B^m \delta A_m = B^m \Gamma^n_{mp} A_n dx^p. \tag{8.18}$$

As the equality (8.18) is valid for an arbitrary contravariant vector B^m, we have

$$\delta A_m = +\Gamma^n_{mp} A_n dx^p. \tag{8.19}$$

On the other hand, analogously to the case of the contravariant vectors, we may write

$$DA_m = dA_m - \delta A_m = \frac{\partial A_m}{\partial x^p} dx^p - \Gamma^n_{mp} A_n dx^p, \tag{8.20}$$

or

$$DA_m = \left(\frac{\partial A_m}{\partial x^p} - \Gamma_{mp}^n A_n \right) dx^p = \frac{DA_m}{Dx^p} dx^p. \tag{8.21}$$

8.2 Covariant Derivatives

8.2.1 Covariant Derivatives of Vectors

From Equation (8.15), we conclude that in the tensor analysis the differential of a contravariant vector dA^m, which does not have a tensor character, is replaced by the tensor differential DA^m. Comparing the results (8.5) and (8.15), we see that, following the same approach, we need to replace the partial derivative of a contravariant vector with respect to a coordinate, which does not have a tensor character either, by the *covariant derivative of a contravariant vector* with respect to a coordinate, as follows:

$$\frac{\partial A^m}{\partial x^p} \rightarrow \frac{DA^m}{Dx^p}, \tag{8.22}$$

where the *covariant derivative of a contravariant vector* is defined by (8.15), as follows:

$$\frac{DA^m}{Dx^p} = \frac{\partial A^m}{\partial x^p} + \Gamma_{np}^m A^n. \tag{8.23}$$

In the Descartes coordinates, where $\Gamma_{np}^m = 0$, the covariant derivative reduces to the corresponding partial derivative.

From the result (8.21), we see that the *covariant derivative of a covariant vector* is defined by the expression

$$\frac{DA_m}{Dx^p} = \frac{\partial A_m}{\partial x^p} - \Gamma_{mp}^n A_n. \tag{8.24}$$

The results (8.23) and (8.24) show that the covariant differentiation of both contravariant and covariant vectors gives the corresponding second-order tensors. In general, by covariant differentiation we preserve the tensor character of an arbitrary tensor, but we increase its order by one. It should be noted that for both contravariant and covariant vectors the order is increased by one additional *covariant index*. Because this operation always increases the number of covariant indices of an arbitrary tensor by one, the corresponding derivative has been named the *covariant derivative*.

8.2.2   Covariant Derivatives of Tensors

Let us consider a special case of a second-order contravariant tensor, obtained as a product of two contravariant vectors A^m and B^m, denoted by $C^{mn} = A^m B^n$. By parallel translation, we obtain

$$\delta(A^m B^n) = A^m \delta B^n + B^n \delta A^m. \tag{8.25}$$

Using the definition (8.11), this expression becomes

$$\delta(A^m B^n) = -A^m \Gamma^n_{kp} B^k dx^p - B^n \Gamma^m_{kp} A^k dx^p$$

$$\delta(A^m B^n) = -\left(\Gamma^n_{kp} A^m B^k + \Gamma^m_{kp} A^k B^n \right) dx^p \tag{8.26}$$

$$\delta C^{mn} = -\left(\Gamma^n_{kp} C^{mk} + \Gamma^m_{kp} C^{kn} \right) dx^p.$$

Because of the linear character of the operation of parallel translation, the result (8.26) is also valid for an arbitrary contravariant tensor C^{mn}. As the differential of a contravariant tensor C^{mn} is by definition given by

$$DC^{mn} = dC^{mn} - \delta C^{mn}, \tag{8.27}$$

substituting from (8.26), we obtain

$$DC^{mn} = \left(\frac{\partial C^{mn}}{\partial x^p} + \Gamma^n_{kp} C^{mk} + \Gamma^m_{kp} C^{kn} \right) dx^p. \tag{8.28}$$

From (8.28), the covariant derivative of a second-order contravariant tensor C^{mn} is defined as

$$\frac{DC^{mn}}{Dx^p} = \frac{\partial C^{mn}}{\partial x^p} + \Gamma^n_{kp} C^{mk} + \Gamma^m_{kp} C^{kn}. \tag{8.29}$$

The same approach can be used for a second-order covariant tensor $C_{mn} = A_m B_n$, where we may write

$$\delta(A_m B_n) = B_n \delta A_m + A_m \delta B_n$$

$$\delta(A_m B_n) = +B_n \Gamma^k_{mp} A_k dx^p + A_m \Gamma^k_{np} B_k dx^p$$

$$\delta(A_m B_n) = \left(\Gamma^k_{mp} A_k B_n + \Gamma^k_{np} A_m B_k \right) dx^p \tag{8.30}$$

$$\delta C_{mn} = \left(\Gamma^k_{mp} C_{kn} + \Gamma^k_{np} C_{mk} \right) dx^p.$$

Again, because of the linear character of the operation of parallel translation, the result (8.30) is also valid for an arbitrary covariant tensor C_{mn}. As the differential of a covariant tensor C_{mn} is by definition given by

$$DC_{mn} = dC_{mn} - \delta C_{mn}, \tag{8.31}$$

substituting from (8.30), we obtain

$$DC_{mn} = \left(\frac{\partial C_{mn}}{\partial x^p} - \Gamma_{mp}^k C_{kn} - \Gamma_{np}^k C_{mk} \right) dx^p. \qquad (8.32)$$

From (8.32), the covariant derivative of a second-order covariant tensor C_{mn} is

$$\frac{DC_{mn}}{Dx^p} = \frac{\partial C_{mn}}{\partial x^p} - \Gamma_{mp}^k C_{kn} - \Gamma_{np}^k C_{mk}. \qquad (8.33)$$

By simple consideration of Equations (8.29) and (8.33), it is easy to define the covariant derivative of a mixed second-order tensor C_n^m as

$$\frac{DC_n^m}{Dx^p} = \frac{\partial C_n^m}{\partial x^p} + \Gamma_{kp}^m C_n^k - \Gamma_{np}^k C_k^m. \qquad (8.34)$$

Analogously to the definitions (8.29), (8.33), and (8.34), it is possible to define covariant derivatives of tensors of an arbitrary order and type. As an example, let us write down the covariant derivative of a mixed third-order tensor C_{nk}^m:

$$\frac{DC_{nk}^m}{Dx^p} = \frac{\partial C_{nk}^m}{\partial x^p} + \Gamma_{sp}^m C_{nk}^s - \Gamma_{np}^s C_{sk}^m - \Gamma_{pk}^s C_{ns}^m. \qquad (8.35)$$

The example (8.35) clearly represents the general method for construction of the covariant derivatives of arbitrary tensors.

8.3 Properties of Covariant Derivatives

There are a number of important rules and special cases that apply to covariant differentiation. These rules and special cases are frequently used in tensor calculations, and the most important ones are listed next.

1. The covariant derivation is a linear operation and the following linearity condition is fulfilled:

$$\frac{D}{Dx^p}(\alpha A^m + \beta B^m) = \alpha \frac{DA^m}{Dx^p} + \beta \frac{DB^m}{Dx^p}. \qquad (8.36)$$

2. The covariant derivative of a product of two tensors is defined in the same way as the usual derivative of a product of two functions. As an example, for the product $A^m C_{mn}$, we may write

$$\frac{D}{Dx^p}(A^m C_{mn}) = \frac{DA^m}{Dx^p} C_{mn} + A^m \frac{DC_{mn}}{Dx^p}. \qquad (8.37)$$

3. The covariant derivative of the metric tensor is equal to zero. In order to prove this proposition, let us observe that for a covariant vector DA_m, in the same way as for any other vector, we have

$$DA_m = g_{mn}DA^n. \tag{8.38}$$

On the other hand, using $A_m = g_{mn}A^n$, we may write

$$DA_m = D(g_{mn}A^n) = g_{mn}DA^n + A^n Dg_{mn}. \tag{8.39}$$

As A^n is an arbitrary vector, by comparison of Equations (8.38) and (8.39) we immediately conclude that

$$Dg_{mn} = 0 \quad \Rightarrow \quad Dg^{mn} = 0. \tag{8.40}$$

In other words, with respect to the operation of covariant differentiation, the metric tensor can be treated as a constant, regardless of the coordinate dependence of its components.

4. The covariant derivative of the δ-symbol is equal to zero. In order to prove this proposition, we use the earlier results

$$Dg_{nk} = 0, \quad Dg^{mk} = 0. \tag{8.41}$$

Using (6.17) and (8.41), we may write

$$D\delta_n^m = D(g^{mk}g_{nk}) = g^{mk}Dg_{nk} + g_{nk}Dg^{mk} = 0. \tag{8.42}$$

Using (8.34), it is also possible to prove this proposition by direct calculation:

$$\frac{D\delta_n^m}{Dx^p} = \frac{\partial \delta_n^m}{\partial x^p} + \Gamma_{kp}^m \delta_n^k - \Gamma_{np}^k \delta_k^m = \Gamma_{np}^m - \Gamma_{np}^m = 0. \tag{8.43}$$

5. The covariant derivative of an invariant scalar function is equal to its ordinary partial derivative, and we may write

$$\frac{D\phi}{Dx^p} = \frac{\partial \phi}{\partial x^p}. \tag{8.44}$$

The partial derivative of a scalar function is therefore transformed as a regular covariant vector. In order to prove this proposition, let us observe a scalar $\phi = A^m B_m$, where A^m is some absolute contravariant vector and B_m is some absolute covariant vector. The covariant derivative of the scalar function ϕ is given by

$$\frac{D\phi}{Dx^p} = \frac{D}{Dx^p}(A^m B_m) = \frac{DA^m}{Dx^p}B_m + A^m \frac{DB_m}{Dx^p}, \tag{8.45}$$

or

$$\frac{D\phi}{Dx^p} = \frac{\partial A^m}{\partial x^p} B_m + \Gamma^m_{kp} A^k B_m + A^m \frac{\partial B_m}{\partial x^p} - \Gamma^k_{mp} A^m B_k. \qquad (8.46)$$

By interchanging the dummy indices $k \leftrightarrow m$, we see that the second term on the right-hand side of Equation (8.46) is cancelled by the fourth term, and we obtain

$$\frac{D\phi}{Dx^p} = \frac{\partial A^m}{\partial x^p} B_m + A^m \frac{\partial B_m}{\partial x^p} = \frac{\partial}{\partial x^p}(A^m B_m) = \frac{\partial \phi}{\partial x^p}, \qquad (8.47)$$

which proves the proposition (8.44).

8.4 Absolute Derivatives of Tensors

Let us assume that a curve C in the N-dimensional metric space is given by means of N parameter equations

$$x^m = x^m(s) \quad (m = 1, 2, \ldots, N), \qquad (8.48)$$

where s is some scalar parameter. Let us now consider a covariant vector A_m defined along the curve C as a function of the parameter s.

The *absolute derivative* of the vector A_m with respect to the parameter s is defined using (8.21) as follows:

$$\frac{DA_m}{ds} = \left(\frac{\partial A_m}{\partial x^p} - \Gamma^n_{mp} A_n \right) \frac{dx^p}{ds}, \qquad (8.49)$$

or

$$\frac{DA_m}{ds} = \frac{DA_m}{Dx^p} \frac{dx^p}{ds}. \qquad (8.50)$$

In the same way we define the absolute derivative of a contravariant vector A^m, as follows:

$$\frac{DA^m}{ds} = \frac{DA^m}{Dx^p} \frac{dx^p}{ds}. \qquad (8.51)$$

The definitions of the absolute derivatives (8.50) and (8.51), for covariant and contravariant vectors, respectively, are easily generalized to the case of an arbitrary tensor. For example, the absolute derivative of a tensor C^m_{nk} is given by

$$\frac{DC^m_{nk}}{ds} = \frac{DC^m_{nk}}{Dx^p} \frac{dx^p}{ds}, \qquad (8.52)$$

where the covariant derivative of the tensor C^m_{nk} is given by Equation (8.35), i.e.,

$$\frac{DC^m_{nk}}{Dx^p} = \frac{\partial C^m_{nk}}{\partial x^p} + \Gamma^m_{sp}C^s_{nk} - \Gamma^s_{np}C^m_{sk} - \Gamma^s_{pk}C^m_{ns}. \tag{8.53}$$

The properties of the absolute derivative are the same as the properties of the covariant derivative and will not be repeated in this section.

▶ Chapter 9

Christoffel Symbols

9.1 Properties of Christoffel Symbols

In order to define various tensor derivatives in the previous chapter, we have introduced a system Γ_{np}^m, called the Christoffel symbol of the second kind. The objective of this chapter is to specify this system and outline its most important properties.

Let us first prove that the Christoffel symbols are symmetric with respect to their lower indices. In order to prove this statement we note that if A_m is an arbitrary covariant vector, then the quantity

$$\frac{DA_m}{Dx^p} - \frac{DA_p}{Dx^m} \tag{9.1}$$

is a second-order covariant tensor. Let us now assume that the vector A_m can be given by the expression

$$A_m = \frac{D\phi}{Dx^m} = \frac{\partial\phi}{\partial x^m}, \tag{9.2}$$

where ϕ is some arbitrary scalar. Then the expression (9.1) can be written in the form

$$\frac{DA_m}{Dx^p} - \frac{DA_p}{Dx^m} = \frac{\partial A_m}{\partial x^p} - \Gamma_{mp}^k A_k - \frac{\partial A_p}{\partial x^m} + \Gamma_{pm}^k A_k, \tag{9.3}$$

or

$$\frac{DA_m}{Dx^p} - \frac{DA_p}{Dx^m}\frac{\partial^2\phi}{\partial x^p\partial x^m} - \frac{\partial^2\phi}{\partial x^m\partial x^p} + \left(\Gamma_{pm}^k - \Gamma_{mp}^k\right)\frac{\partial\phi}{\partial x^k}. \qquad (9.4)$$

The first and second term in Equation (9.4) cancel each other and we obtain

$$\frac{DA_m}{Dx^p} - \frac{DA_p}{Dx^m} = \left(\Gamma_{pm}^k - \Gamma_{mp}^k\right)\frac{\partial\phi}{\partial x^k}. \qquad (9.5)$$

As the left-hand side of Equation (9.5) is a second-order covariant tensor, we conclude that the right-hand side of Equation (9.5), i.e.,

$$\left(\Gamma_{pm}^k - \Gamma_{mp}^k\right)\frac{\partial\phi}{\partial x^k}, \qquad (9.6)$$

is also a second-order covariant tensor. However, we know by definition of the Christoffel symbols that they are identically equal to zero in all points of the Euclidean metric space described by the Descartes coordinates. Thus the second-order covariant tensor (9.6) is identically equal to zero in the Descartes coordinates. But a tensor that is identically equal to zero in one coordinate system must be equal to zero in any other coordinate system. Thus we conclude that

$$\frac{DA_m}{Dx^p} - \frac{DA_p}{Dx^m} = \left(\Gamma_{pm}^k - \Gamma_{mp}^k\right)\frac{\partial\phi}{\partial x^k} = 0. \qquad (9.7)$$

As the vector A_m, given by

$$A_m = \frac{D\phi}{Dx^m} = \frac{\partial\phi}{\partial x^m}, \qquad (9.8)$$

is an arbitrary covariant vector and ϕ is an arbitrary scalar function, we conclude from Equation (9.7) that

$$\Gamma_{pm}^k = \Gamma_{mp}^k \quad (k,m,p = 1,2,\ldots,N), \qquad (9.9)$$

and the Christoffel symbols of the second kind are indeed symmetric with respect to their lower indices. From (9.9) it is evident that, for the Christoffel symbols of the first kind, we have

$$\Gamma_{n,pm} = \Gamma_{n,mp} \quad (m,n,p = 1,2,\ldots,N). \qquad (9.10)$$

Let us now derive the transformation law of Christoffel symbols with respect to the transformations of coordinates. If a contravariant vector is defined with respect to the systems of coordinates $\{x^k\}$ and $\{z^k\}$ by N coordinates A^m and $\bar{A}^m$, respectively, then it transforms according to the transformation law

$$\bar{A}^m = \frac{\partial z^m}{\partial x^j} A^j. \tag{9.11}$$

Using (9.11) we may write

$$\frac{\partial \bar{A}^m}{\partial z^p} = \frac{\partial}{\partial z^p}\left(\frac{\partial z^m}{\partial x^j} A^j\right) = \frac{\partial x^k}{\partial z^p}\frac{\partial}{\partial x^k}\left(\frac{\partial z^m}{\partial x^j} A^j\right), \tag{9.12}$$

or, after calculating the derivative in the parentheses,

$$\frac{\partial \bar{A}^m}{\partial z^p} = \frac{\partial x^k}{\partial z^p}\frac{\partial z^m}{\partial x^j}\frac{\partial A^j}{\partial x^k} + \frac{\partial x^k}{\partial z^p}\frac{\partial^2 z^m}{\partial x^k \partial x^j} A^j. \tag{9.13}$$

The transformation law (9.13) is just a direct confirmation of the fact that the partial derivative of a contravariant vector is not a tensor, as we have shown indirectly in the previous chapter. On the other hand, the covariant derivative of the contravariant vector is a mixed second-order tensor and it transforms according to the transformation law

$$\frac{D\bar{A}^m}{Dz^p} = \frac{\partial x^k}{\partial z^p}\frac{\partial z^m}{\partial x^j}\frac{DA^j}{Dx^k}. \tag{9.14}$$

Here we recall the definitions of covariant derivatives of contravariant vectors (8.23), in the coordinate system $\{z^k\}$,

$$\frac{D\bar{A}^m}{Dz^p} = \frac{\partial \bar{A}^m}{\partial z^p} + \bar{\Gamma}^m_{np}\bar{A}^n, \tag{9.15}$$

and in the coordinate system $\{x^k\}$, i.e.,

$$\frac{DA^j}{Dx^k} = \frac{\partial A^j}{\partial x^k} + \Gamma^j_{lk}A^l. \tag{9.16}$$

Substituting (9.15) and (9.16) into (9.14), we obtain

$$\frac{\partial \bar{A}^m}{\partial z^p} + \bar{\Gamma}^m_{np}\bar{A}^n = \frac{\partial x^k}{\partial z^p}\frac{\partial z^m}{\partial x^j}\frac{\partial A^j}{\partial x^k} + \frac{\partial x^k}{\partial z^p}\frac{\partial z^m}{\partial x^j}\Gamma^j_{lk}A^l. \tag{9.17}$$

Substituting (9.13) into (9.17) we obtain

$$\frac{\partial x^k}{\partial z^p}\frac{\partial z^m}{\partial x^j}\frac{\partial A^j}{\partial x^k} + \frac{\partial x^k}{\partial z^p}\frac{\partial^2 z^m}{\partial x^k \partial x^j}A^j + \bar{\Gamma}^m_{np}\bar{A}^n$$
$$= \frac{\partial x^k}{\partial z^p}\frac{\partial z^m}{\partial x^j}\frac{\partial A^j}{\partial x^k} + \frac{\partial x^k}{\partial z^p}\frac{\partial z^m}{\partial x^j}\Gamma^j_{lk}A^l. \tag{9.18}$$

The first term on the left-hand side cancels the first term on the right-hand side in Equation (9.18), and by using

$$A^j = \delta^j_l A^l, \quad \bar{A}^n = \frac{\partial z^n}{\partial x^l} A^l, \tag{9.19}$$

we obtain from (9.18)

$$\left(\delta^j_l \frac{\partial x^k}{\partial z^p} \frac{\partial^2 z^m}{\partial x^k \partial x^j} + \bar{\Gamma}^m_{np} \frac{\partial z^n}{\partial x^l} \right) A^l = \frac{\partial x^k}{\partial z^p} \frac{\partial z^m}{\partial x^j} \Gamma^j_{lk} A^l. \tag{9.20}$$

As Equation (9.20) is valid for an arbitrary contravariant vector A^l, this tensor can be omitted. After omitting A^l and multiplying the remaining equation by

$$\frac{\partial z^p}{\partial x^k} \frac{\partial x^j}{\partial z^m}, \tag{9.21}$$

we finally obtain

$$\Gamma^j_{lk} = \frac{\partial x^j}{\partial z^m} \frac{\partial z^n}{\partial x^l} \frac{\partial z^p}{\partial x^k} \bar{\Gamma}^m_{np} + \frac{\partial x^j}{\partial z^m} \frac{\partial^2 z^m}{\partial x^k \partial x^l}. \tag{9.22}$$

Equation (9.22) is the required transformation law for the Christoffel symbols, and it shows that the system Γ^j_{lk} behaves as a tensor only with respect to the linear transformations of Descartes coordinate systems, when we have

$$\frac{\partial^2 z^m}{\partial x^k \partial x^l} = 0. \tag{9.23}$$

In general, as we have concluded before, the Christoffel symbol is not a tensor.

9.2   Relation to the Metric Tensor

For practical calculation of the Christoffel symbol we need to find its relation to the metric tensor. In order to find this relation, we start with the identity

$$\frac{D g_{mn}}{dx^p} = \frac{\partial g_{mn}}{\partial x^p} - \Gamma^k_{mp} g_{kn} - \Gamma^k_{np} g_{mk} = 0, \tag{9.24}$$

or

$$\frac{\partial g_{mn}}{\partial x^p} = \Gamma_{n,mp} + \Gamma_{m,np}. \tag{9.25}$$

The permutations of indices in Equation (9.25) give the equation

$$\frac{\partial g_{pm}}{\partial x^n} = \Gamma_{m,pn} + \Gamma_{p,mn} = 0, \tag{9.26}$$

and, with reversed signs, the equation

$$-\frac{\partial g_{np}}{\partial x^m} = -\Gamma_{p,nm} - \Gamma_{n,pm} = 0. \tag{9.27}$$

By adding together Equations (9.25), (9.26), and (9.27) and using the symmetry property (9.10), we obtain

$$\Gamma_{m,np} = \frac{1}{2}\left(\frac{\partial g_{mn}}{\partial x^p} + \frac{\partial g_{pm}}{\partial x^n} - \frac{\partial g_{np}}{\partial x^m}\right). \tag{9.28}$$

The result (9.28) is a definition of the Christoffel symbol of the first kind as a function of the metric tensor. The Christoffel symbol of the second kind is then given by

$$\Gamma_{np}^m = g^{mk}\Gamma_{k,np} = \frac{1}{2}g^{mk}\left(\frac{\partial g_{kn}}{\partial x^p} + \frac{\partial g_{pk}}{\partial x^n} - \frac{\partial g_{np}}{\partial x^k}\right). \tag{9.29}$$

The result (9.29) is a definition of the Christoffel symbol of the second kind as a function of the metric tensor. Using the definition (9.29), we can calculate the contracted Christoffel symbol of the second kind Γ_{nm}^m, as follows:

$$\Gamma_{nm}^m = \frac{1}{2}\left(g^{mk}\frac{\partial g_{kn}}{\partial x^m} + g^{mk}\frac{\partial g_{mk}}{\partial x^n} - g^{mk}\frac{\partial g_{nm}}{\partial x^k}\right). \tag{9.30}$$

Because of the symmetry of the metric tensor and the symmetry of the Christoffel symbol with respect to the interchange of its lower indices, the first term in the parentheses of Equation (9.30) cancels the third term, and we obtain

$$\Gamma_{nm}^m = \frac{1}{2}g^{mk}\frac{\partial g_{mk}}{\partial x^n}. \tag{9.31}$$

The determinant of the metric tensor is given by the expression (6.12) as follows:

$$g = g_{mk}G^{mk}, \tag{9.32}$$

where G^{mk} is a cofactor of the determinant $|g_{mn}|$ that corresponds to the element g_{mn}. The differential of the determinant g is, in view of Equation (9.32), equal to

$$dg = \frac{\partial g}{\partial g_{mk}}dg_{mk} = G^{mk}dg_{mk}, \tag{9.33}$$

Using now the definition of the contravariant metric tensor (6.16), i.e.,

$$G^{mk} = gg^{mk}, \tag{9.34}$$

in Equation (9.33), we obtain

$$dg = gg^{mk} dg_{mk} \quad \Rightarrow \quad \frac{\partial g}{\partial x^n} = gg^{mk} \frac{\partial g_{mk}}{\partial x^n}. \tag{9.35}$$

Substituting (9.35) into the result (9.31), we obtain

$$\Gamma^m_{nm} = \frac{1}{2g} \frac{\partial g}{\partial x^n} = \frac{\partial \ln \sqrt{g}}{\partial x^n}. \tag{9.36}$$

If we now recall the result (6.17) in the form

$$g^{mk} g_{nk} = \delta^m_n, \tag{9.37}$$

we may contract it to obtain

$$g^{mk} g_{mk} = \delta^m_m = N \quad (k, m = 1, 2, \dots, N). \tag{9.38}$$

Thus by differentiation with respect to the coordinate x^n, we have

$$\frac{\partial}{\partial x^n} (g^{mk} g_{mk}) = g^{mk} \frac{\partial g_{mk}}{\partial x^n} + g_{mk} \frac{\partial g^{mk}}{\partial x^n} = 0, \tag{9.39}$$

or

$$g^{mk} \frac{\partial g_{mk}}{\partial x^n} = -g_{mk} \frac{\partial g^{mk}}{\partial x^n}. \tag{9.40}$$

Thus there is an alternative expression for the contracted Christoffel symbol of the second kind obtained by substituting (9.40) into (9.31):

$$\Gamma^m_{nm} = -\frac{1}{2} g_{mk} \frac{\partial g^{mk}}{\partial x^n}. \tag{9.41}$$

It is also of interest to calculate the quantity $g^{kn} \Gamma^m_{kn}$, since it frequently appears in various calculations in the tensor analysis. This quantity is, by definition, given by

$$g^{kn} \Gamma^m_{kn} = g^{kn} g^{ml} \Gamma_{l,kn} = \frac{1}{2} g^{kn} g^{ml} \left(\frac{\partial g_{lk}}{\partial x^n} + \frac{\partial g_{nl}}{\partial x^k} - \frac{\partial g_{kn}}{\partial x^l} \right). \tag{9.42}$$

Using the symmetry of the metric tensor and interchanging the dummy indices $k \leftrightarrow n$, we see that the first two terms in the parentheses of Equation (9.42) are equal to each other. Thus we may rewrite (9.42) as follows:

$$g^{kn} \Gamma^m_{kn} = g^{kn} g^{ml} \left(\frac{\partial g_{kl}}{\partial x^n} - \frac{1}{2} \frac{\partial g_{kn}}{\partial x^l} \right). \tag{9.43}$$

The first term on the right-hand side of Equation (9.43) can be rewritten as follows:

$$g^{kn} g^{ml} \frac{\partial g_{kl}}{\partial x^n} = g^{kn} \frac{\partial}{\partial x^n} \left(g^{ml} g_{kl} \right) - g^{kn} g_{kl} \frac{\partial g^{ml}}{\partial x^n}$$

$$= g^{kn} \frac{\partial}{\partial x^n} \left(\delta^m_k \right) - \delta^n_l \frac{\partial g^{ml}}{\partial x^n} = -\frac{\partial g^{mn}}{\partial x^n}. \tag{9.44}$$

Combining the results (9.31) and (9.36), we may write

$$\frac{1}{2} g^{kn} \frac{\partial g_{kn}}{\partial x^l} = \frac{1}{2g} \frac{\partial g}{\partial x^l} = \frac{\partial \ln \sqrt{g}}{\partial x^l}, \tag{9.45}$$

and the second term on the right-hand side of Equation (9.43) becomes

$$-\frac{1}{2} g^{ml} g^{kn} \frac{\partial g_{kn}}{\partial x^l} = -g^{ml} \frac{\partial \ln \sqrt{g}}{\partial x^l} = -g^{mn} \frac{1}{\sqrt{g}} \frac{\partial \sqrt{g}}{\partial x^n}. \tag{9.46}$$

Substituting (9.44) and (9.46) into (9.43) we obtain

$$g^{kn} \Gamma^m_{kn} = -\left(\frac{\partial g^{mn}}{\partial x^n} + g^{mn} \frac{1}{\sqrt{g}} \frac{\partial \sqrt{g}}{\partial x^n} \right)$$

$$= -\frac{1}{\sqrt{g}} \left(\sqrt{g} \frac{\partial g^{mn}}{\partial x^n} + g^{mn} \frac{\partial \sqrt{g}}{\partial x^n} \right), \tag{9.47}$$

and using the rule for the derivative of the product $\sqrt{g} g^{mn}$, we finally obtain the expression for the quantity $g^{kn} \Gamma^m_{kn}$ as follows:

$$g^{kn} \Gamma^m_{kn} = -\frac{1}{\sqrt{g}} \frac{\partial}{\partial x^n} \left(\sqrt{g} g^{mn} \right). \tag{9.48}$$

► **Chapter 10**

Differential Operators

10.1   The Hamiltonian ∇-Operator

In the ordinary three-dimensional Euclidean metric space the Hamiltonian ∇-operator is defined by the expression

$$\nabla_m = \partial_m = \frac{\partial}{\partial x^m} \quad (m = 1, 2, 3). \tag{10.1}$$

However, by acting with this operator on an arbitrary vector or tensor we obtain a system that does not have the tensor character, as shown by, e.g., Equation (9.13). In an arbitrary N-dimensional generalized curvilinear coordinate system, instead of ∇_m, we therefore use the covariant Hamiltonian operator D_m, defined by

$$D_m = \frac{D}{Dx^m} \quad (m = 1, 2, \dots, N). \tag{10.2}$$

Using the operator D_m, we define the operators called *gradient, divergence, curl,* and *Laplacian* in the following four sections.

10.2   Gradient of Scalars

The *gradient* of a scalar function ϕ is a covariant vector with its covariant coordinates defined as follows:

$$D_m\phi = \frac{D\phi}{Dx^m} = \frac{\partial\phi}{\partial x^m}. \tag{10.3}$$

79

The contravariant coordinates of the gradient of a scalar ϕ are obtained as follows:

$$D^m\phi = g^{mn}\frac{D\phi}{Dx^n} = g^{mn}\frac{\partial\phi}{\partial x^n}. \tag{10.4}$$

In the orthogonal curvilinear coordinates, with the metric form given by (6.60), i.e.,

$$ds^2 = (h_1 dx^1)^2 + (h_2 dx^2)^2 + (h_3 dx^3)^2, \tag{10.5}$$

the physical coordinates of the gradient of a scalar ϕ are given by

$$D_{(m)}\phi = \frac{1}{h_M}D_m\phi = \frac{1}{h_M}\frac{\partial\phi}{\partial x^m} \quad (m, M = 1, 2, 3). \tag{10.6}$$

The definition (10.6) gives the correct physical coordinates of the gradient in the well-known cases of cylindrical and spherical coordinate systems. As an illustration, let us consider the spherical coordinates where, using (6.65) and (6.66), we may write

$$\begin{array}{ccc} x^1 = r, & x^2 = \theta, & x^3 = \varphi \\ h_1 = 1, & h_2 = r, & h_3 = r\sin\theta. \end{array} \tag{10.7}$$

The gradient of a scalar function ϕ is then, according to (10.6), given by

$$\text{grad } \phi = \frac{\partial\phi}{\partial r}\vec{e}_r + \frac{1}{r}\frac{\partial\phi}{\partial\theta}\vec{e}_\theta + \frac{1}{r\sin\theta}\frac{\partial\phi}{\partial\varphi}\vec{e}_\varphi, \tag{10.8}$$

where $\vec{e}_r$, $\vec{e}_\theta$, and $\vec{e}_\varphi$ are the unit vectors in spherical coordinates.

10.3   Divergence of Vectors and Tensors

The *divergence* of an arbitrary contravariant vector A^m is a scalar obtained by the composition of the covariant Hamiltonian operator D_m with that contravariant vector A^m, i.e.,

$$D_m A^m = \frac{\partial A^m}{\partial x^m} + \Gamma^m_{nm}A^n. \tag{10.9}$$

Using here the result (9.36), we may write

$$D_m A^m = \frac{\partial A^m}{\partial x^m} + \frac{\partial\ln\sqrt{g}}{\partial x^n}A^n. \tag{10.10}$$

Since n is a dummy index in the second term on the right-hand side of Equation (10.10), it can be changed to m. Thus we may write

$$D_m A^m = \frac{\partial A^m}{\partial x^m} + \frac{1}{\sqrt{g}} \frac{\partial \sqrt{g}}{\partial x^m} A^m$$

$$D_m A^m = \frac{1}{\sqrt{g}} \left(\sqrt{g} \frac{\partial A^m}{\partial x^m} + \frac{\partial \sqrt{g}}{\partial x^m} A^m \right). \tag{10.11}$$

Finally, the expression for the divergence of the contravariant vector A^m is given by

$$D_m A^m = \frac{1}{\sqrt{g}} \frac{\partial}{\partial x^m} \left(\sqrt{g} A^m \right). \tag{10.12}$$

If a vector is defined by its covariant coordinates A_m, then the divergence of such a vector is written in the form

$$D^m A_m = g^{mn} D_n A_m = D_n \left(g^{mn} A_m \right), \tag{10.13}$$

where we have used the property that the metric tensor behaves as a scalar with respect to covariant differentiation. In such a case, using (10.12), we obtain

$$D^m A_m = \frac{1}{\sqrt{g}} \frac{\partial}{\partial x^m} \left(\sqrt{g} g^{mn} A_n \right). \tag{10.14}$$

Analogously to the case of a contravariant vector (10.12), it is possible to define the divergence of an arbitrary tensor with respect to one of its upper indices. Thus, the divergence of the tensor T_p^{mn} is defined by

$$D_m T_p^{mn} = \frac{1}{\sqrt{g}} \frac{\partial}{\partial x^m} \left(\sqrt{g} T_p^{mn} \right). \tag{10.15}$$

In the orthogonal curvilinear coordinates, with the metric form given by (10.5), we have $\sqrt{g} = h_1 h_2 h_3$, and the physical coordinates of the divergence of a contravariant vector A^m are given by

$$D_m A^m = \frac{1}{h_1 h_2 h_3} \frac{\partial}{\partial x^m} \left(h_1 h_2 h_3 \frac{A_{(m)}}{h_M} \right), \tag{10.16}$$

where the physical coordinates of the contravariant vector A^m are given by

$$A_{(m)} = h_M A^m \quad (m, M = 1, 2, 3). \tag{10.17}$$

The definition (10.16) gives the correct physical expressions for the divergence in the well-known cases of cylindrical and spherical coordinate systems. As an illustration, let us again consider the spherical coordinates defined by (10.7). The physical expression for the divergence of the vector $\vec{A}$ in the spherical coordinates is given by

$$\text{div } \vec{A} = \frac{1}{r^2}\frac{\partial}{\partial r}\left(r^2 A_r\right) + \frac{1}{r\sin\theta}\frac{\partial}{\partial\theta}\left(\sin\theta A_\theta\right) + \frac{1}{r\sin\theta}\frac{\partial A_\varphi}{\partial\theta}. \qquad (10.18)$$

10.4   Curl of Vectors

In an arbitrary N-dimensional metric space the curl of a vector function A_m is a second-order covariant tensor F_{mn}, defined by the expression

$$F_{mn} = D_m A_n - D_n A_m \quad (m, n = 1, 2, \ldots, N). \qquad (10.19)$$

In the special case of a three-dimensional metric space, it is possible to construct a contravariant vector C^k related to the tensor (10.19) using the three-dimensional totally antisymmetric Ricci tensor ε^{kmn}, defined according to the general definition (6.23) as follows:

$$\varepsilon^{kmn} = \frac{1}{\sqrt{g}}e^{kmn} \quad (k, m, n = 1, 2, 3). \qquad (10.20)$$

Thus the curl in the three-dimensional metric space is defined by

$$C^k = \frac{1}{2}\varepsilon^{kmn}F_{mn} = \varepsilon^{kmn}D_m A_n = \frac{1}{\sqrt{g}}e^{kmn}D_m A_n. \qquad (10.21)$$

In the orthogonal curvilinear coordinates, with the metric form given by (10.5), we have $\sqrt{g} = h_1 h_2 h_3$, and the physical coordinates of the curl of a covariant vector A_n are given by

$$\text{curl}_{(k)}\,\vec{A} = h_K C^k = \frac{h_K}{h_1 h_2 h_3}e^{kmn}\frac{\partial}{\partial x^m}\left(h_N A_{(n)}\right) \qquad (10.22)$$

where the physical coordinates of the covariant vector A_n are given by

$$A_{(n)} = \frac{A_n}{h_N} \quad (n, N = 1, 2, 3). \qquad (10.23)$$

As the expression (10.22) is somewhat more complex and generally not readily understood, we expand it for the three components of the curl in three dimensions:

$$\text{curl}_{(1)}\, \vec{A} = \frac{1}{h_2 h_3} \left(\frac{\partial (h_3 A_{(3)})}{\partial x^2} - \frac{\partial (h_2 A_{(2)})}{\partial x^3} \right)$$

$$\text{curl}_{(2)}\, \vec{A} = \frac{1}{h_1 h_3} \left(\frac{\partial (h_1 A_{(1)})}{\partial x^3} - \frac{\partial (h_3 A_{(3)})}{\partial x^1} \right) \tag{10.24}$$

$$\text{curl}_{(3)}\, \vec{A} = \frac{1}{h_1 h_2} \left(\frac{\partial (h_2 A_{(2)})}{\partial x^1} - \frac{\partial (h_1 A_{(1)})}{\partial x^2} \right).$$

The expressions (10.24) are usually structured into a determinant defined as follows:

$$\text{curl}\, \vec{A} = \frac{1}{h_1 h_2 h_3} \begin{vmatrix} h_1 \vec{1} & h_2 \vec{2} & h_3 \vec{3} \\ \frac{\partial}{\partial x^1} & \frac{\partial}{\partial x^2} & \frac{\partial}{\partial x^3} \\ h_1 A_{(1)} & h_2 A_{(2)} & h_3 A_{(3)} \end{vmatrix}. \tag{10.25}$$

The definition (10.25) gives the correct physical expressions for the components of the curl of a vector function in the well-known cases of cylindrical and spherical coordinate systems. As an illustration, we can write down this determinant for the spherical coordinates defined by (10.7). Thus we may write

$$\text{curl}\, \vec{A} = \frac{1}{r^2 \sin \theta} \begin{vmatrix} \vec{e}_r & r\,\vec{e}_\theta & r \sin \theta\, \vec{e}_\varphi \\ \frac{\partial}{\partial r} & \frac{\partial}{\partial \theta} & \frac{\partial}{\partial \varphi} \\ A_r & r A_\theta & r \sin \theta\, A_\varphi \end{vmatrix}, \tag{10.26}$$

or

$$\text{curl}\, \vec{A} = \frac{1}{r \sin \theta} \left[\frac{\partial}{\partial \theta} \left(\sin \theta\, A_\varphi \right) - \frac{\partial A_\theta}{\partial \varphi} \right] \vec{e}_r$$

$$+ \frac{1}{r} \left[\frac{1}{\sin \theta} \frac{\partial A_r}{\partial \varphi} - \frac{\partial}{\partial r} \left(r A_\varphi \right) \right] \vec{e}_\theta + \frac{1}{r} \left[\frac{\partial}{\partial r} \left(r A_\theta \right) - \frac{\partial A_r}{\partial \theta} \right] \vec{e}_\varphi. \tag{10.27}$$

10.5 Laplacian of Scalars and Tensors

The *Laplacian* of a scalar function in the three-dimensional vector analysis is defined by the expression

$$\Delta \phi = \text{div}\, (\text{grad}\, \phi). \tag{10.28}$$

In the tensor analysis this expression becomes

$$\Delta\phi = D_m D^m \phi = D_m \left(g^{mn} D_n \phi\right) = D_m \left(g^{mn} \frac{\partial\phi}{\partial x^n}\right). \tag{10.29}$$

From (10.29) we see that the Laplacian of the scalar function ϕ is a divergence of a vector given by its covariant coordinates

$$D_n\phi = \frac{\partial\phi}{\partial x^n}. \tag{10.30}$$

Using here the definition (10.14) of the divergence of a vector defined by its covariant coordinates, we may write

$$\Delta\phi = \frac{1}{\sqrt{g}} \frac{\partial}{\partial x^m} \left(\sqrt{g}g^{mn} \frac{\partial\phi}{\partial x^n}\right). \tag{10.31}$$

In the orthogonal curvilinear coordinates, with the metric form given by (10.5), we have $\sqrt{g} = h_1 h_2 h_3$ and the Laplacian of the scalar function ϕ becomes

$$\Delta\phi = \frac{1}{h_1 h_2 h_3} \frac{\partial}{\partial x^m} \left(h_1 h_2 h_3 g^{mn} \frac{\partial\phi}{\partial x^n}\right), \tag{10.32}$$

where

$$[g^{mn}] = \begin{bmatrix} (h_1)^{-2} & 0 & 0 \\ 0 & (h_2)^{-2} & 0 \\ 0 & 0 & (h_3)^{-2} \end{bmatrix}. \tag{10.33}$$

Substituting (10.33) into (10.32) we obtain

$$\Delta\phi = \frac{1}{h_1 h_2 h_3} \left[\frac{\partial}{\partial x^1} \left(\frac{h_2 h_3}{h_1} \frac{\partial\phi}{\partial x^1}\right) \right.$$
$$\left. + \frac{\partial}{\partial x^2} \left(\frac{h_1 h_3}{h_2} \frac{\partial\phi}{\partial x^2}\right) + \frac{\partial}{\partial x^3} \left(\frac{h_1 h_2}{h_3} \frac{\partial\phi}{\partial x^3}\right) \right]. \tag{10.34}$$

The definition (10.34) gives the correct physical expressions for the Laplacian in the well-known cases of cylindrical and spherical coordinate systems. As an illustration, in the spherical coordinates defined by (10.7), the expression for the Laplacian becomes

$$\Delta\phi = \frac{1}{r^2} \frac{\partial}{\partial r} \left(r^2 \frac{\partial\phi}{\partial r}\right) + \frac{1}{r^2 \sin\theta} \frac{\partial}{\partial\theta} \left(\sin\theta \frac{\partial\phi}{\partial\theta}\right) + \frac{1}{r^2 \sin^2\theta} \frac{\partial^2\phi}{\partial\varphi^2}. \tag{10.35}$$

10.6 Integral Theorems for Tensor Fields

In the three-dimensional vector analysis there are two important integral theorems, called the Stokes theorem and the Gauss theorem. These theorems remain valid in the tensor analysis and their formulation is generalized in such a way that they can be applied to integrals in arbitrary N-dimensional metric spaces. These theorems in the tensor notation will be discussed in the rest of this section.

10.6.1 Stokes Theorem

In the usual three-dimensional vector notation the *Stokes theorem* is defined as

$$\oint_C \vec{A} \cdot d\vec{r} = \int_S \text{curl } \vec{A} \cdot d\vec{S}. \qquad (10.36)$$

The integral on the right-hand side of Equation (10.36) is a surface integral over a surface S bound by a closed contour C. The integral on the left-hand side of Equation (10.36) is a line integral round the closed contour C running along the boundaries of the surface S. In the tensor notation the formulation of the Stokes theorem has the following form:

$$\oint_C A_m dx^m = \frac{1}{2} \int_S F_{mn} dS^{mn}, \qquad (10.37)$$

where F_{mn} is the *curl tensor* of the vector A_m, defined by Equation (10.19), i.e.,

$$F_{mn} = D_m A_n - D_n A_m. \qquad (10.38)$$

On the other hand,

$$dS^{mn} = dx^m dx^n \qquad (10.39)$$

is the contravariant tensor of an infinitesimal element of the surface S.

From the definition (10.37) we see that the Stokes theorem in tensor notation is valid for arbitrary generalized metric spaces. Let us now prove that Equation (10.37) is equivalent to Equation (10.36) in the three-dimensional metric space. We first note that in the three-dimensional space we have

$$A_m dx^m = \vec{A} \cdot d\vec{x} \quad (m = 1, 2, 3). \qquad (10.40)$$

From (10.40) we see that the integrals on the left-hand sides of Equations (10.36) and (10.37) are indeed equivalent to each other. In order to prove that the right-hand sides of Equations (10.36) and (10.37) are also equivalent to each other, it is convenient to introduce a three-dimensional covariant *surface vector* as follows:

$$dS_k = \frac{1}{2}\varepsilon_{kmn}dx^m dx^n = \frac{1}{2}\sqrt{g}e_{kmn}dx^m dx^n. \qquad (10.41)$$

In the Descartes coordinates, where $\sqrt{g} = 1$, the components of this vector are given by

$$dS_1 = dx^2 dx^3, \quad dS_2 = dx^1 dx^3, \quad dS_3 = dx^1 dx^2. \qquad (10.42)$$

Furthermore, analogously to (10.21), it is convenient to introduce the contravariant *curl vector*, denoted by C^k and related to the tensor F_{mn} by the expression

$$C^k = \frac{1}{2}\varepsilon^{kjl}F_{jl} = \varepsilon^{kjl}D_j A_l = \frac{1}{\sqrt{g}}e^{kjl}D_j A_l = \left(\mathrm{curl}\,\vec{A}\right)^k. \qquad (10.43)$$

Using (10.43) we may write

$$\mathrm{curl}\,\vec{A}\cdot d\vec{S} = \left(\mathrm{curl}\,\vec{A}\right)^k dS_k = \tfrac{1}{2}e^{kjl}e_{kmn}D_j A_l dx^m dx^n. \qquad (10.44)$$

On the other hand, the e-symbols satisfy the identity

$$e^{kjl}e_{kmn} = \delta_m^j\delta_n^l - \delta_n^j\delta_m^l. \qquad (10.45)$$

Substituting (10.45) into (10.44) we obtain

$$\mathrm{curl}\,\vec{A}\cdot d\vec{S} = \tfrac{1}{2}\left(\delta_m^j\delta_n^l - \delta_n^j\delta_m^l\right)D_j A_l dx^m dx^n$$
$$\mathrm{curl}\,\vec{A}\cdot d\vec{S} = \tfrac{1}{2}\left(D_m A_n - D_n A_m\right)dx^m dx^n \qquad (10.46)$$
$$\mathrm{curl}\,\vec{A}\cdot d\vec{S} = \tfrac{1}{2}F_{mn}dS^{mn}.$$

From (10.46) we see that the integrals on the right-hand sides of Equations (10.36) and (10.37) are also equivalent to each other in the three-dimensional metric space.

10.6.2 Gauss Theorem

In the usual three-dimensional vector notation the *Gauss theorem* is defined by means of the equation

$$\oint_S \vec{A}\cdot d\vec{S} = \int_\Omega \mathrm{div}\,\vec{A}\,d\Omega. \qquad (10.47)$$

The integral on the right-hand side of Equation (10.47) is a volume integral over a volume Ω bound by a closed surface S. The integral on the left-hand side of Equation (10.47) is a surface integral over the closed surface S enclosing the volume Ω. The volume element $d\Omega$ is a relative scalar with the weight $M = -1$. If the volume element is defined with respect to the systems of coordinates $\{x^k\}$ and $\{z^k\}$ by $d\Omega$ and $d\bar{\Omega}^m$, respectively, then according to the Jacobi theorem it transforms as follows:

$$d\bar{\Omega} = \left| \frac{\partial z^m}{\partial x^k} \right| d\Omega = \Delta^{-1} d\Omega. \tag{10.48}$$

Therefore, in the tensor formulation of the Gauss theorem, we use the invariant volume element $\sqrt{g}d\Omega$ to obtain

$$\oint_S A^m dS_m = \int_\Omega D_m A^m \sqrt{g}\, d\Omega. \tag{10.49}$$

The tensor formulation of the Gauss theorem (10.48) is valid for arbitrary generalized metric spaces. The equivalence of Equations (10.47) and (10.48) in the Descartes coordinates, where $\sqrt{g} = 1$ and $D_m = \nabla_m$, is evident.

The Gauss theorem can be extended to the case of an arbitrary tensor with at least one upper index. As an example, for a mixed third-order tensor T_k^{mn}, the Gauss theorem is given by

$$\oint_S T_k^{mn} dS_m = \int_\Omega D_m T_k^{mn} \sqrt{g}\, d\Omega. \tag{10.50}$$

Chapter 11

Geodesic Lines

In Euclidean metric spaces the shortest distance between two given points is a straight line. However, even the simplest example of a two-dimensional Riemannian space on the surface of a unit sphere shows clearly that in the Riemannian spaces there are in general no straight lines. It is therefore of interest to find the curves of minimum (or at least stationary) arc length connecting the two given points in a Riemannian space.

In Riemannian spaces, we in general do not impose the requirement to find the lines of minimum arc length, and we replace it by the requirement to find lines with a stationary arc length between the two given points. We can illustrate this on the simple example of the two-dimensional Riemannian space on the surface of a unit sphere. For both arcs on the same circle, connecting the two given points on the unit sphere, we can only say that their length is of a stationary character, but we *cannot* say that they are of minimum length.

The curves with stationary arc length between two given points A and B are called the *geodesic lines*, and they are determined by the solutions of the corresponding *geodesic differential equations*. In order to construct these differential equations, we will use the variational calculus and *Lagrange equations*, which will be derived in the next section.

11.1 Lagrange Equations

Let us observe two fixed points A and B in an arbitrary metric space. Their coordinates are given by the parameter equations

$$x^m = x^m(S_A), \quad x^m = x^m(S_B), \qquad (11.1)$$

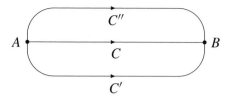

Figure 3. Curves connecting the fixed points A and B

where s is the arc-length parameter. Between the fixed points A and B we can draw a family of curves of various arc lengths, as shown in Figure 3.

Within the family of curves, which connect the two fixed points A and B as shown in Figure 3, there is a single curve C with a stationary arc length, which we call the geodesic line. All other curves between the two fixed points A and B, e.g., C' and C'', do not have a stationary arc length and they deviate from the geodesic lines by some variations $\delta x^m(s)$ for all values of the parameter s. It should be noted that points A and B are fixed and common for all the curves connecting these two points. Therefore, the variations δx^m at these two points are equal to zero:

$$\delta x^m(S_A) = \delta x^m(S_B) = 0. \tag{11.2}$$

Let us now define a function,

$$L = L\left(x^m, \frac{dx^m}{ds}, s\right), \tag{11.3}$$

along each of the lines connecting the points A and B. This function may depend on the coordinates x^m and their first derivatives with respect to the parameter s, as well as on the parameter s itself, as indicated in the definition (11.3). This function is usually called the *Lagrange function* or *Lagrangian*. The integral of this function between the points A and B is given by

$$I = \int_{S_A}^{S_B} L\left(x^m, \frac{dx^m}{ds}, s\right) ds, \tag{11.4}$$

and it is usually called the *action integral*. According to the Hamiltonian variational principle, the integral (11.4) has a stationary value along the geodesic line C. Using the variational principle, we can derive the differential equation satisfied by the Lagrangian function (11.3). The variation of the action integral (11.4) along the geodesic line C is equal to zero, and we may write

$$\delta I = \int_{S_A}^{S_B} \delta L\left(x^m, \frac{dx^m}{ds}, s\right) ds = 0. \tag{11.5}$$

By definition of the variation we have

$$\delta L = L\left(x^m + \delta x^m, \frac{dx^m}{ds} + \delta\frac{dx^m}{ds}, s\right) - L\left(x^m, \frac{dx^m}{ds}, s\right). \tag{11.6}$$

Since the variations of the coordinates and their derivatives with respect to the parameter are small, we can expand the first term on the right-hand side of Equation (11.6) into a Taylor series and keep only the zeroth- and first-order terms. Thus we obtain

$$\delta L = L\left(x^m, \frac{dx^m}{ds}, s\right) + \frac{\partial L}{\partial x^m}\delta x^m + \frac{\partial L}{\partial\left(\frac{dx^m}{ds}\right)}\delta\frac{dx^m}{ds} - L\left(x^m, \frac{dx^m}{ds}, s\right)$$

$$= \frac{\partial L}{\partial x^m}\delta x^m + \frac{\partial L}{\partial\left(\frac{dx^m}{ds}\right)}\delta\frac{dx^m}{ds}. \tag{11.7}$$

Since the differentiation with respect to the parameter is independent from the variation, the order of these two operations can be interchanged. Thus we may write

$$\delta L = \frac{\partial L}{\partial x^m}\delta x^m + \frac{\partial L}{\partial\left(\frac{dx^m}{ds}\right)}\frac{d}{ds}\left(\delta x^m\right), \tag{11.8}$$

or

$$\delta L = \frac{\partial L}{\partial x^m}\delta x^m - \frac{d}{ds}\left[\frac{\partial L}{\partial\left(\frac{dx^m}{ds}\right)}\right]\delta x^m ds + \frac{d}{ds}\left[\frac{\partial L}{\partial\left(\frac{dx^m}{ds}\right)}\delta x^m\right]. \tag{11.9}$$

Substituting (11.9) into (11.5) we obtain

$$\delta I = \int_{S_A}^{S_B}\left\{\frac{\partial L}{\partial x^m} - \frac{d}{ds}\left[\frac{\partial L}{\partial\left(\frac{dx^m}{ds}\right)}\right]\right\}\delta x^m ds + \frac{\partial L}{\partial\left(\frac{dx^m}{ds}\right)}\delta x^m\Bigg|_{S_A}^{S_B} = 0. \tag{11.10}$$

The second term of Equation (11.10), which is calculated at the boundary points of the geodesic lines A and B, vanishes due to the condition (11.2). Thus we obtain

$$\delta I = \int_{S_A}^{S_B}\left\{\frac{\partial L}{\partial x^m} - \frac{d}{ds}\left[\frac{\partial L}{\partial\left(\frac{dx^m}{ds}\right)}\right]\right\}\delta x^m ds = 0. \tag{11.11}$$

Since the variations δx^m are arbitrary and in general different from zero, the result (11.11) requires that the following equations be satisfied:

$$\frac{\partial L}{\partial x^m} - \frac{d}{ds}\left[\frac{\partial L}{\partial\left(\frac{dx^m}{ds}\right)}\right] = 0 \quad (m = 1, 2, \ldots, N) \qquad (11.12)$$

The system of N Equations (11.12) is a system of differential equations that must be satisfied by the Lagrangian function (11.3) along the geodesic line C. These equations are usually called the *Lagrange equations*.

11.2 Geodesic Equations

In order to construct the geodesic equations that define the curve with a stationary arc length, we may choose the arc length itself as the action integral with zero variation. Using the expression (6.38) we write

$$I = \int_{S_A}^{S_B} \sqrt{g_{kn}\frac{dx^k}{ds}\frac{dx^n}{ds}}\,ds. \qquad (11.13)$$

From (11.13) we see that the Lagrangian function is given by the expression

$$L = \sqrt{g_{kn}\frac{dx^k}{ds}\frac{dx^n}{ds}}, \qquad (11.14)$$

and it is equal to unity along the geodesic line C, where

$$ds^2 = g_{kn}dx^k dx^n \quad (k, n = 1, 2, \ldots, N). \qquad (11.15)$$

Along the geodesic line, where the metric condition (11.15) is fulfilled, we have

$$\frac{\partial L}{\partial\left(\frac{dx^m}{ds}\right)} = \frac{1}{2}\left(g_{kn}\frac{dx^k}{ds}\frac{dx^n}{ds}\right)^{-\frac{1}{2}}2g_{ml}\frac{dx^l}{ds} = g_{ml}\frac{dx^l}{ds}, \qquad (11.16)$$

and

$$\frac{d}{ds}\left[\frac{\partial L}{\partial\left(\frac{dx^m}{ds}\right)}\right] = \frac{d}{ds}\left(g_{ml}\frac{dx^l}{ds}\right) = g_{ml}\frac{d^2x^l}{ds^2} + \frac{\partial g_{ml}}{\partial x^k}\frac{dx^l}{ds}\frac{dx^k}{ds}. \qquad (11.17)$$

By interchanging the dummy indices $k \leftrightarrow l$, we may rewrite the second term on the right-hand side of the result (11.17) as follows:

$$\frac{\partial g_{ml}}{\partial x^k} \frac{dx^l}{ds} \frac{dx^k}{ds} = \frac{1}{2} \left(\frac{\partial g_{ml}}{\partial x^k} + \frac{\partial g_{mk}}{\partial x^l} \right) \frac{dx^l}{ds} \frac{dx^k}{ds}. \tag{11.18}$$

Substituting (11.18) into (11.17) we obtain

$$\frac{d}{ds} \left[\frac{\partial L}{\partial \left(\frac{dx^m}{ds} \right)} \right] = g_{ml} \frac{d^2 x^l}{ds^2} + \frac{1}{2} \left(\frac{\partial g_{ml}}{\partial x^k} + \frac{\partial g_{mk}}{\partial x^l} \right) \frac{dx^l}{ds} \frac{dx^k}{ds}. \tag{11.19}$$

On the other hand we have

$$\frac{\partial L}{\partial x^m} = \frac{1}{2} \left(g_{kn} \frac{dx^k}{ds} \frac{dx^n}{ds} \right)^{-\frac{1}{2}} \frac{\partial g_{lk}}{\partial x^m} \frac{dx^l}{ds} \frac{dx^k}{ds}. \tag{11.20}$$

Using the metric condition (11.15), we obtain

$$\frac{\partial L}{\partial x^m} = \frac{1}{2} \frac{\partial g_{lk}}{\partial x^m} \frac{dx^l}{ds} \frac{dx^k}{ds}. \tag{11.21}$$

Substituting (11.19) and (11.21) into the Lagrange Equations (11.12), we obtain

$$g_{ml} \frac{d^2 x^l}{ds^2} + \frac{1}{2} \left(\frac{\partial g_{ml}}{\partial x^k} + \frac{\partial g_{mk}}{\partial x^l} - \frac{\partial g_{lk}}{\partial x^m} \right) \frac{dx^l}{ds} \frac{dx^k}{ds} = 0. \tag{11.22}$$

Now, using the definition of the Christoffel symbols of the first kind (9.28), i.e.,

$$\Gamma_{m,lk} = \frac{1}{2} \left(\frac{\partial g_{ml}}{\partial x^k} + \frac{\partial g_{km}}{\partial x^l} - \frac{\partial g_{lk}}{\partial x^m} \right), \tag{11.23}$$

we obtain from (11.22)

$$g_{ml} \frac{d^2 x^l}{ds^2} + \Gamma_{m,lk} \frac{dx^l}{ds} \frac{dx^k}{ds} = 0. \tag{11.24}$$

The composition of Equation (11.24) with the contravariant metric tensor g^{mn} gives the most commonly used formulation of the *geodesic equations*, as follows:

$$\frac{d^2 x^n}{ds^2} + \Gamma^n_{lk} \frac{dx^l}{ds} \frac{dx^k}{ds} = 0. \tag{11.25}$$

If a set of parameter equations of some curve C in a generalized N-dimensional metric space is given by

$$x^m = x^m(s) \quad (m = 1, 2, \ldots, N), \tag{11.26}$$

then the tangent vector to this curve is defined by

$$u^m = \frac{dx^m}{ds} \quad (m = 1, 2, \ldots, N). \tag{11.27}$$

Using (11.27), Equation (11.25) can be written in the form

$$\frac{du^n}{ds} + \Gamma^n_{lk} u^l u^k = 0. \tag{11.28}$$

Using Equation (11.22) we can also write

$$g_{ml} \frac{du^l}{ds} + \frac{1}{2}\left(\frac{\partial g_{ml}}{\partial x^k} + \frac{\partial g_{mk}}{\partial x^l}\right) u^l u^k - \frac{1}{2}\frac{\partial g_{lk}}{\partial x^m} u^l u^k = 0. \tag{11.29}$$

Interchanging the dummy indices $l \leftrightarrow k$ in the second term on the left-hand side of Equation (11.29), we obtain

$$g_{ml} \frac{du^l}{ds} + \frac{\partial g_{ml}}{\partial x^k} u^l u^k - \frac{1}{2}\frac{\partial g_{lk}}{\partial x^m} u^l u^k = 0. \tag{11.30}$$

On the other hand, using the definition $u_m = g_{ml} u^l$, we may write

$$\frac{du_m}{ds} = \frac{d}{ds}\left(g_{ml} u^l\right) = g_{ml}\frac{du^l}{ds} + \frac{dg_{ml}}{ds} u^l = g_{ml}\frac{du^l}{ds} + \frac{\partial g_{ml}}{\partial x^k} u^l u^k. \tag{11.31}$$

Substituting (11.31) into (11.30), we obtain

$$\frac{du_m}{ds} - \frac{1}{2}\frac{\partial g_{lk}}{\partial x^m} u^l u^k = 0, \tag{11.32}$$

or

$$\frac{d^2 x_m}{ds^2} - \frac{1}{2}\frac{\partial g_{lk}}{\partial x^m}\frac{dx^l}{ds}\frac{dx^k}{ds} = 0. \tag{11.33}$$

The equation (11.33) is an alternative form of the geodesic equations, which does not explicitly involve the Christoffel symbols. Using the geodesic Equations (11.25), we may also write

$$\frac{du^n}{ds} + \Gamma^n_{lk} u^l \frac{dx^k}{ds} = \left(\frac{\partial u^n}{\partial x^k} + \Gamma^n_{lk} u^l \right) \frac{dx^k}{ds} = 0. \tag{11.34}$$

or, using the definition of the covariant derivative of a contravariant vector (8.23),

$$\frac{Du^n}{Dx^k} = \frac{\partial u^n}{\partial x^k} + \Gamma^n_{lk} u^l, \tag{11.35}$$

we find that the absolute derivative of the tangent vector (11.27) with respect to the parameter s is equal to zero:

$$\frac{Du^n}{ds} = 0. \tag{11.36}$$

This means that if we translate the vector u^m along the geodesic line from the point x^m to the point $x^m + dx^m$, it will coincide with the tangent vector $u^m + du^m$ at the point $x^m + dx^m$. This is a specific property of the tangent vector u^m along the geodesic line. In the Descartes coordinates, where $\Gamma^n_{lk} = 0$, the geodesic differential equation becomes

$$\frac{d^2 y^n}{ds^2} = 0. \tag{11.37}$$

The solution of this equation is a set of linear equations

$$y^n = a^n s + b^n, \tag{11.38}$$

which represent the parameter equations of a straight line, which confirms our earlier statement that the geodesic lines in the Euclidean metric spaces are straight lines.

► Chapter 12

The Curvature Tensor

12.1 │ Definition of the Curvature Tensor

In the Riemannian space the parallel translation of a vector between two given points is a path-dependent operation, i.e., it gives different results along different paths. In particular, if a vector is parallelly translated along a closed path back to the same point of origin, it will not coincide with the original vector. Let us therefore derive a general formula for the change of a vector after a parallel translation along some infinitesimally small closed contour C. This change, denoted by ΔA_m, is obtained using the result (8.19), in the following manner:

$$\Delta A_m = \oint_C \delta A_m = \oint_C \Gamma^n_{mp} A_n dx^p. \tag{12.1}$$

Using the Stokes theorem, this line integral over the closed contour C can be transformed into a surface integral over a surface S bound by the closed contour C, i.e.,

$$\Delta A_m = \frac{1}{2} \oint_C \left[D_k \left(\Gamma^n_{mp} A_n \right) - D_p \left(\Gamma^n_{mk} A_n \right) \right] dS^{kp}, \tag{12.2}$$

where we use the covariant operator

$$D_k = \frac{D}{Dx^k}. \tag{12.3}$$

Let us also, for the sake of simplicity, introduce the non-tensor operator

$$\partial_k = \frac{\partial}{\partial x^k}. \tag{12.4}$$

Using the definition of the covariant derivative of a second-order covariant tensor (8.33), we may then write

$$D_k C_{mp} = \partial_k C_{mp} - \Gamma^l_{mk} C_{lp} - \Gamma^l_{pk} C_{ml}$$
$$D_p C_{mk} = \partial_p C_{mk} - \Gamma^l_{mp} C_{lk} - \Gamma^l_{kp} C_{ml}. \tag{12.5}$$

Subtracting the two Equations (12.5) from each other we obtain

$$D_k C_{mp} - D_p C_{mk}$$
$$= \partial_k C_{mp} - \Gamma^l_{mk} C_{lp} - \Gamma^l_{pk} C_{ml} - \partial_p C_{mk} + \Gamma^l_{mp} C_{lk} + \Gamma^l_{kp} C_{ml}$$
$$= \partial_k C_{mp} - \partial_p C_{mk} + \Gamma^l_{mp} C_{lk} - \Gamma^l_{mk} C_{lp} + \Gamma^l_{kp} C_{ml} - \Gamma^l_{pk} C_{ml}. \tag{12.6}$$

Using here the symmetry of the Christoffel symbols with respect to their lower indices, we see that the last two terms in Equation (12.6) cancel each other. Thus, using the symmetric surface tensor $dS^{kp} = dS^{pk}$, we may write

$$\left(D_k C_{mp} - D_p C_{mk} \right) dS^{kp}$$
$$= \left(\partial_k C_{mp} - \partial_p C_{mk} + \Gamma^l_{mp} C_{lk} - \Gamma^l_{mk} C_{lp} \right) dS^{kp}. \tag{12.7}$$

By interchanging the indices $k \leftrightarrow p$ in one of the last two terms in parentheses and using the symmetry of the surface tensor dS^{kp}, we see that these two terms cancel each other. The result (12.7) therefore becomes

$$(D_k C_{mp} - D_p C_{mk}) dS^{kp} = (\partial_k C_{mp} - \partial_p C_{mk}) dS^{kp} \tag{12.8}$$

Applying (12.8) with $C_{mp} = \Gamma^n_{mp} A_n$ in Equation (12.2), we obtain

$$\Delta A_m = \frac{1}{2} \oint_C \left[\partial_k \left(\Gamma^n_{mp} A_n \right) - \partial_p \left(\Gamma^n_{mk} A_n \right) \right] dS^{kp}, \tag{12.9}$$

or

$$\Delta A_m = \frac{1}{2} \oint_C \left(\partial_k \Gamma^n_{mp} A_n - \partial_p \Gamma^n_{mk} A_n + \Gamma^n_{mp} \partial_k A_n - \Gamma^n_{mk} \partial_p A_n \right) dS^{kp}. \tag{12.10}$$

On the other hand, the change of the vector A_n along the contour C is due to the parallel translation (8.19) and is given by

$$\delta A_n = \Gamma^l_{np} A_l dx^p \quad \Rightarrow \quad \partial_p A_n = \Gamma^l_{np} A_l. \tag{12.11}$$

Substituting (12.11) into (12.10), we obtain

$$\Delta A_m = \frac{1}{2} \oint_C \left(\partial_k \Gamma_{mp}^n A_n - \partial_p \Gamma_{mk}^n A_n + \Gamma_{mp}^n \Gamma_{nk}^l A_l - \Gamma_{mk}^n \Gamma_{np}^l A_l \right) dS^{kp}.$$
$$(12.12)$$

As the labels of the dummy indices are irrelevant, we can interchange the indices $n \leftrightarrow l$ in the last two terms of the integral (12.12). Thus we obtain

$$\Delta A_m = \frac{1}{2} \oint_C \left(\partial_k \Gamma_{mp}^n - \partial_p \Gamma_{mk}^n + \Gamma_{mp}^l \Gamma_{lk}^n - \Gamma_{mk}^l \Gamma_{lp}^n \right) A_n dS^{kp} \quad (12.13)$$

Introducing here the notation

$$R_{mkp}^n = \partial_k \Gamma_{mp}^n - \partial_p \Gamma_{mk}^n + \Gamma_{mp}^l \Gamma_{lk}^n - \Gamma_{mk}^l \Gamma_{lp}^n, \quad (12.14)$$

the result (12.13) becomes

$$\Delta A_m = \frac{1}{2} \oint_C R_{mkp}^n A_n dS^{kp}. \quad (12.15)$$

Since the closed contour C is infinitesimally small, it is possible to replace the integrand of the integral (12.15) by its value at some point enclosed by the contour C, and to bring it outside the integral. Thus, we finally obtain a general formula for the change of a vector after a parallel translation along some infinitesimally small closed contour C in the form

$$\Delta A_m = \frac{1}{2} R_{mkp}^n A_n \Delta S^{kp}. \quad (12.16)$$

The mixed fourth-order tensor defined by the expression (12.14) is called the *curvature tensor* of a given metric space. The tensor character of the curvature tensor is evident from (12.16), since A_n is a covariant vector, ΔS^{kp} is a contravariant surface tensor, and ΔA_m is a difference of two covariant vectors at the same point on the contour C. The curvature tensor plays the key role in the theory of gravitational field and is very important in tensor analysis in general.

The formula, analogous to (12.16), for a contravariant vector A^m can be obtained using the fact that a scalar $A^m B_m$ is invariant under parallel translations. Thus we may write

$$\Delta(A^m B_m) = \Delta A^m B_m + A^m \Delta B_m$$
$$= \Delta A^n B_n + A^m \Delta B_m = 0. \quad (12.17)$$

Using here the result (12.16) we obtain

$$\Delta A^n B_n + A^m \frac{1}{2} R^n_{mkp} B_n \Delta S^{kp} = \left(\Delta A^n + \frac{1}{2} R^n_{mkp} A^m \Delta S^{kp} \right) B_n = 0.$$

$$(12.18)$$

As the covariant vector B_n is arbitrary, we obtain the formula

$$\Delta A^n = -\frac{1}{2} R^n_{mkp} A^m \Delta S^{kp}.$$

$$(12.19)$$

In a given Euclidean space the curvature tensor is identically equal to zero, since it is possible to choose the coordinates where $\Gamma^n_{mp} \equiv 0$ and therefore $R^n_{mkp} = 0$, in the entire metric space. Because of its tensor character, the curvature tensor is then equal to zero in any other coordinate system defined in the Euclidean metric space. It is related to the observation that, in the Euclidean metric space, the parallel translation is not a path-dependent operation and the parallel translation along a closed curve C does not change the translated vector.

The reverse argument is valid as well. If the curvature tensor is equal to zero in the entire metric space, i.e., $R^n_{mkp} = 0$, then such a space is Euclidean. Indeed, in any metric space, it is possible to construct a local Descartes system in an infinitesimally small portion of that space. On the other hand, if $R^n_{mkp} = 0$ in the entire space, then the parallel translation is a unique path-independent operation by which this infinitesimally small portion of a given space can be translated to any other portion of that space. Thus we may construct Descartes coordinates in the entire space and the given space is Euclidean.

As a summary, the *criterion for determining the character of a space* is that a metric space described by a metric tensor g_{mn} is Euclidean if and only if $R^n_{mkp} = 0$.

12.2 Properties of the Curvature Tensor

From the definition of the curvature tensor

$$R^n_{mkp} = \partial_k \Gamma^n_{mp} - \partial_p \Gamma^n_{mk} + \Gamma^l_{mp} \Gamma^n_{lk} - \Gamma^l_{mk} \Gamma^n_{lp},$$

$$(12.20)$$

it is easily seen that it is antisymmetric with respect to the interchange of the last two lower indices $k \leftrightarrow p$, such that we have

$$R^n_{mkp} = -R^n_{mpk}.$$

$$(12.21)$$

It is also possible to show that the curvature tensor satisfies the following identity:

$$R^n_{mkp} + R^n_{pmk} + R^n_{kpm} = 0. \tag{12.22}$$

In order to prove the cyclic identity (12.22), let us rewrite (12.20) as follows:

$$
\begin{aligned}
R^n_{mkp} &= \partial_k \Gamma^n_{mp} - \partial_p \Gamma^n_{mk} + \Gamma^l_{mp} \Gamma^n_{lk} - \Gamma^l_{mk} \Gamma^n_{lp} \\
R^n_{pmk} &= \partial_m \Gamma^n_{pk} - \partial_k \Gamma^n_{pm} + \Gamma^l_{pk} \Gamma^n_{lm} - \Gamma^l_{pm} \Gamma^n_{lk} \\
R^n_{kpm} &= \partial_p \Gamma^n_{km} - \partial_m \Gamma^n_{kp} + \Gamma^l_{km} \Gamma^n_{lp} - \Gamma^l_{kp} \Gamma^n_{lm}.
\end{aligned}
\tag{12.23}
$$

By inspection of Equations (12.23), using the symmetry of the Christoffel symbols with respect to the interchange of the two lower indices, we see that for each term in these three equations there is a counterterm with the opposite sign. Thus, by adding together the three equations (12.23) we obtain the cyclic identity (12.22). In addition to the mixed curvature tensor (12.20), it is sometimes useful to define the covariant curvature tensor by

$$R_{mnkp} = g_{ml} R^l_{nkp}, \tag{12.24}$$

or, using the definition (12.20),

$$R_{mnkp} = g_{ml} \partial_k \Gamma^l_{np} - g_{ml} \partial_p \Gamma^l_{nk} + \Gamma^l_{np} \Gamma_{m,lk} - \Gamma^l_{nk} \Gamma_{m,lp}. \tag{12.25}$$

In order to obtain a more symmetric form of the covariant curvature tensor (12.24), we use Equation (9.25) in the form

$$
\begin{aligned}
\partial_p g_{ml} &= \Gamma_{l,mp} + \Gamma_{m,lp} \\
\partial_k g_{ml} &= \Gamma_{l,mk} + \Gamma_{m,lk}
\end{aligned}
\tag{12.26}
$$

and the definition of the Christoffel symbol of the first kind (9.28) in the form

$$\Gamma_{m,np} = \frac{1}{2}(\partial_p g_{mn} + \partial_n g_{pm} - \partial_m g_{np}). \tag{12.27}$$

Using (12.26) and (12.27) we can calculate

$$
\begin{aligned}
g_{ml} \partial_k \Gamma^l_{np} &= g_{ml} \partial_k \left(g^{ls} \Gamma_{s,np} \right) = \left(g_{ml} \partial_k g^{ls} \right) \Gamma_{s,np} + \delta^s_m \partial_k \Gamma_{s,np} \\
&= - \left(g^{ls} \partial_k g_{ml} \right) \Gamma_{s,np} + \partial_k \Gamma_{m,np} = -\Gamma^l_{np} \partial_k g_{ml} + \partial_k \Gamma_{m,np} \\
&= -\Gamma^l_{np} \left(\Gamma_{l,mk} + \Gamma_{m,lk} \right) + \frac{1}{2} \left(\partial_k \partial_p g_{mn} + \partial_k \partial_n g_{pm} - \partial_k \partial_m g_{np} \right).
\end{aligned}
\tag{12.28}
$$

By interchanging the indices $k \leftrightarrow p$ in (12.28) we also obtain

$$g_{ml}\partial_p\Gamma^l_{nk} = -\Gamma^l_{nk}\left(\Gamma_{l,mp} + \Gamma_{m,lp}\right)$$

$$+ \frac{1}{2}\left(\partial_p\partial_k g_{mn} + \partial_p\partial_n g_{km} - \partial_p\partial_m g_{nk}\right). \qquad (12.29)$$

Substituting (12.28) and (12.29) into (12.25) we obtain

$$R_{mnkp} = -\Gamma^l_{np}\left(\Gamma_{l,mk} + \Gamma_{m,lk}\right) + \Gamma^l_{nk}\left(\Gamma_{l,mp} + \Gamma_{m,lp}\right)$$

$$+ \frac{1}{2}\left(\partial_k\partial_p g_{mn} + \partial_k\partial_n g_{pm} - \partial_k\partial_m g_{np}\right)$$

$$- \frac{1}{2}\left(\partial_p\partial_k g_{mn} + \partial_p\partial_n g_{km} - \partial_p\partial_m g_{nk}\right)$$

$$+ \Gamma^l_{np}\Gamma_{m,lk} - \Gamma^l_{nk}\Gamma_{m,lp}, \qquad (12.30)$$

or

$$R_{mnkp} = \frac{1}{2}(\partial_k\partial_n g_{pm} + \partial_p\partial_m g_{nk} - \partial_k\partial_m g_{np} - \partial_p\partial_n g_{km})$$

$$+ \Gamma^l_{nk}\Gamma_{l,mp} - \Gamma^l_{np}\Gamma_{l,mk}. \qquad (12.31)$$

Finally we obtain the alternative expression for the covariant curvature tensor in the form

$$R_{mnkp} = \frac{1}{2}\left(\partial_k\partial_n g_{pm} + \partial_p\partial_m g_{nk} - \partial_k\partial_m g_{np} - \partial_p\partial_n g_{km}\right)$$

$$+ g_{lj}\left(\Gamma^l_{nk}\Gamma^j_{mp} - \Gamma^l_{np}\Gamma^j_{mk}\right). \qquad (12.32)$$

From the expression (12.32) we see that the covariant curvature tensor is antisymmetric with respect to the interchange of both the first two indices $(m \leftrightarrow n)$ and the last two indices $(k \leftrightarrow p)$. Thus we have

$$R_{mnkp} = -R_{nmkp}, \quad R_{mnkp} = -R_{mnpk}. \qquad (12.33)$$

From the expression (12.32) we also see that the covariant curvature tensor is symmetric with respect to the interchange of the first two indices and the last two indices $(mn \leftrightarrow kp)$, i.e.,

$$R_{mnkp} = R_{kpmn}. \qquad (12.34)$$

Furthermore we note that the cyclic property (12.22) also remains valid after the contraction with the covariant metric tensor, such that we have

$$R_{mnkp} + R_{mpnk} + R_{mkpn} = 0. \qquad (12.35)$$

Using (12.33) and (12.34), it can also be shown that the cyclic properties analogous to (12.35) are valid not only for the last three, but also for any three indices of the covariant curvature tensor.

12.3 Commutator of Covariant Derivatives

In the Descartes coordinates, with $D_k = \partial_k$, it is allowed to change the order of the covariant differentiation with respect to the two different coordinates. In other words, we may write

$$\partial_k \partial_m A^p = \partial_m \partial_k A^p \quad \Rightarrow \quad D_k D_m A^p = D_m D_k A^p \tag{12.36}$$

or

$$(D_k D_m - D_m D_k)A^p = (\partial_k \partial_m - \partial_m \partial_k)A^p = 0, \tag{12.37}$$

where A^p is an arbitrary contravariant vector. Since A^p is an arbitrary contravariant vector, it can be omitted and we can write the corresponding operator expression as follows:

$$D_k D_m - D_m D_k = [D_m, D_k] = [\partial_m, \partial_k] = 0. \tag{12.38}$$

The quantity $[D_m, D_k]$, introduced in (12.38), is called the *commutator* of the two operators D_m and D_k. This commutator is equal to zero in the Descartes coordinates. Let us now calculate the commutator $[D_m, D_k]$ in the arbitrary curvilinear coordinates. In order to calculate this commutator, we first calculate

$$D_k D_m A^p = D_k(\partial_m A^p + \Gamma^p_{mn} A^n)$$
$$D_k D_m A^p = \partial_k(\partial_m A^p + \Gamma^p_{mn} A^n)$$
$$+ \Gamma^l_{kl}(\partial_m A^l + \Gamma^l_{mn} A^n) + \Gamma^l_{km}(\partial_l A^p + \Gamma^p_{ln} A^n) \tag{12.39}$$
$$D_k D_m A^p = \partial_k \partial_m A^p + \Gamma^p_{mn} \partial_k A^n + A^n \partial_k \Gamma^p_{mn}$$
$$+ \Gamma^p_{kl} \partial_m A^l + \Gamma^p_{kl} \Gamma^l_{mn} A^n - \Gamma^l_{km} \partial_l A^p - \Gamma^l_{km} \Gamma^p_{ln} A^n.$$

Analogously, we may write

$$D_m D_k A^p = \partial_m \partial_k A^p + \Gamma^p_{kn} \partial_m A^n + A^n \partial_m \Gamma^p_{kn} + \Gamma^p_{ml} \partial_k A^l$$
$$+ \Gamma^p_{ml} \Gamma^l_{kn} A^n - \Gamma^l_{mk} \partial_l A^p - \Gamma^l_{mk} \Gamma^p_{ln} A^n. \tag{12.40}$$

If we now form the difference between the expressions (12.39) and (12.40), we see that several terms cancel each other, i.e., the first term cancels the first term, the second term cancels the fourth term, the fourth term cancels the second term, the sixth term cancels the sixth term, and the seventh term cancels the seventh term. Thus we obtain

$$(D_k D_m - D_m D_k)A^p = \left(\partial_k \Gamma^p_{mn} - \partial_m \Gamma^p_{kn} + \Gamma^p_{kl}\Gamma^l_{mn} - \Gamma^p_{ml}\Gamma^l_{nk} \right) A^n. \tag{12.41}$$

Now, substituting the definition of the curvature tensor (12.14), i.e.,

$$R^p_{nkm} = \partial_k \Gamma^p_{mn} - \partial_m \Gamma^p_{kn} + \Gamma^l_{mn}\Gamma^p_{kl} - \Gamma^l_{nk}\Gamma^p_{ml}, \tag{12.42}$$

into Equation (12.41), we obtain

$$(D_k D_m - D_m D_k)A^p = R^p_{nkm}A^n. \tag{12.43}$$

In line with the criterion for determining the character of a space, discussed earlier, we conclude from (12.43) that the change of the order of the covariant differentiation is allowed only in the Euclidean metric spaces with the zero curvature tensor, i.e., with $R^p_{nkm} = 0$.

12.4   Ricci Tensor and Scalar

From the curvature tensor (12.14), by contraction of the single contravariant index with the second covariant index, it is possible to construct a covariant second-order tensor, called the *Ricci tensor*, as follows:

$$R_{mn} = R^p_{mnp} = -R^p_{mpn}. \tag{12.44}$$

Using here the definition of the curvature tensor (12.14) we may write

$$R_{mn} = \partial_n \Gamma^p_{pm} - \partial_p \Gamma^p_{mn} + \Gamma^l_{mp}\Gamma^p_{nl} - \Gamma^p_{mn}\Gamma^l_{lp}. \tag{12.45}$$

It is easily verified that the Ricci tensor can only be defined as in (12.44). Let us for example consider the alternative contracted tensor R^k_{kmn}, which can be written as follows:

$$R^k_{kmn} = \delta^k_l R^l_{kmn} = g^{pk}g_{pl}R^l_{kmn} = g^{pk}R_{pkmn} = 0. \tag{12.46}$$

From the definition (12.46) we see that R^k_{kmn} is a scalar product of the symmetric metric tensor g^{pk} and the antisymmetric covariant curvature tensor R_{pkmn}, which is by definition equal to zero. Using the symmetry properties (12.33) and (12.34) in the definition of the Ricci tensor

$$R_{mn} = R^p_{mnp} = g^{pk}R_{kmnp}, \tag{12.47}$$

and interchanging the dummy indices $k \leftrightarrow p$, we may write

$$R_{mn} = g^{pk} R_{npkm} = g^{kp} R_{nkpm} = g^{pk} R_{knmp} = R_{nm}. \qquad (12.48)$$

Thus the Ricci tensor is symmetric with respect to its two indices, i.e.,

$$R_{mn} = R_{nm} \quad (m, n = 1, 2, \ldots, N). \qquad (12.49)$$

Using the Ricci tensor (12.44), we can define the *Ricci scalar* as follows:

$$R = g^{mn} R_{mn} = g^{mn} g^{pk} R_{kmnp}. \qquad (12.50)$$

The Ricci tensor and Ricci scalar are extensively used in the general theory of relativity and in cosmology.

12.5 Curvature Tensor Components

The components of the curvature tensor are related to each other by means of the symmetry relations (12.33)–(12.35). Thus the components of the curvature tensor are not all independent, and the number of the independent components of the curvature tensor in the N-dimensional space is equal to

$$n = \frac{N^2(N^2 - 1)}{12}. \qquad (12.51)$$

As this general result for the N-dimensional metric space will not be extensively used and its derivation involves long and complex combinatorial manipulations, it will not be considered in detail here. We will only consider the two simplest special cases, a two-dimensional metric space ($N = 2$) and a three-dimensional metric space ($N = 3$).

In a two-dimensional metric space, the indices of the covariant curvature tensor R_{kmnp}, i.e., (k, m, n, p) can assume the values 1 and 2. Thus the components of the covariant curvature tensor can be presented by the following rectangular scheme:

$$[R_{kmnp}] =$$

	$np = 11$	$np = 12$	$np = 21$	$np = 22$
$km = 11$	0	0	0	0
$km = 12$	0	R_{1212}	$-R_{1212}$	0
$km = 21$	0	$-R_{1212}$	R_{1212}	0
$km = 22$	0	0	0	0

(12.52)

From the scheme (12.52) we see that out of possible $2^4 = 16$ components, the 12 components with $k = m$ or $n = p$ or both are equal to zero because of the antisymmetry of the covariant curvature tensor. The remaining four nonzero components with $k \neq m$ and $n \neq p$ are related by antisymmetry relations as well. Thus, according to the expression (12.51), we see that in a two-dimensional metric space there is just one independent component of the covariant curvature tensor, i.e.,

$$n = \frac{4(4-1)}{12} = 1. \tag{12.53}$$

The Ricci scalar is then defined by

$$R = g^{mn} g^{pk} R_{kmnp} = 2\, g^{11} g^{22} R_{1212} - 2\, g^{12} g^{21} R_{1212}$$

$$R = 2\left(g^{11}g^{22} - g^{12}g^{21}\right) R_{1212} = 2\left|g^{mn}\right| R_{1212} = \frac{2}{g} R_{1212}. \tag{12.54}$$

The quantity $R/2$ is equal to the Gauss curvature of a surface, or the inverse of the product of the main radii of the curvature.

Let us now consider a special case of the surface of the unit sphere, with the metric of the form

$$ds^2 = d\theta^2 + \sin^2\theta\, d\varphi^2 \tag{12.55}$$

or

$$ds^2 = (dx^1)^2 + (\sin x^1)^2 (dx^2)^2. \tag{12.56}$$

Thus, the covariant metric tensor can be written in the matrix form

$$[g_{mn}] = \begin{bmatrix} 1 & 0 \\ 0 & (\sin x^1)^2 \end{bmatrix}, \tag{12.57}$$

and the contravariant metric tensor can be written in the matrix form

$$[g^{mn}] = \begin{bmatrix} 1 & 0 \\ 0 & (\sin x^1)^{-2} \end{bmatrix}. \tag{12.58}$$

In this case, the quantity $R/2$ should be equal to unity. In order to show that, we need to calculate the only independent component of the curvature tensor, i.e.,

$$R_{1212} = -\frac{1}{2}\partial_1\partial_1 g_{22} + g_{11}(\Gamma^1_{12}\Gamma^1_{12} - \Gamma^1_{11}\Gamma^1_{22})$$
$$+ g_{22}(\Gamma^2_{12}\Gamma^2_{12} - \Gamma^2_{11}\Gamma^2_{22}). \tag{12.59}$$

On the other hand, the Christoffel symbols of the first kind are defined by

$$\Gamma_{m,np} = \frac{1}{2}(\partial_p g_{mn} + \partial_n g_{pm} - \partial_m g_{np}). \tag{12.60}$$

Substituting (12.56) into (12.60), we obtain

$$\Gamma_{1,11} = 0$$
$$\Gamma_{1,12} = \Gamma_{1,21} = 0$$
$$\Gamma_{1,22} = -\frac{1}{2}\partial_1 g_{22} = -\sin x^1 \cos x^1$$
$$\Gamma_{2,11} = 0$$
$$\Gamma_{2,12} = \Gamma_{2,21} = +\frac{1}{2}\partial_1 g_{22} = +\sin x^1 \cos x^1$$
$$\Gamma_{2,22} = 0$$

(12.61)

The Christoffel symbols of the second kind are defined as

$$\Gamma_{np}^m = g^{ml}\Gamma_{l,np},$$ (12.62)

and in our example, they are given by

$$\Gamma_{11}^1 = g^{11}\Gamma_{1,11} + g^{12}\Gamma_{2,11} = 0$$
$$\Gamma_{12}^1 = \Gamma_{21}^1 = g^{11}\Gamma_{1,21} + g^{12}\Gamma_{2,21} = 0$$
$$\Gamma_{22}^1 = g^{11}\Gamma_{1,22} + g^{12}\Gamma_{2,22} = -\sin x^1 \cos x^1$$
$$\Gamma_{11}^2 = g^{21}\Gamma_{1,11} + g^{22}\Gamma_{2,11} = 0$$
$$\Gamma_{12}^2 = \Gamma_{21}^2 = g^{21}\Gamma_{1,12} + g^{22}\Gamma_{2,12} = +(\sin x^1)^{-1}\cos x^1$$
$$\Gamma_{22}^2 = g^{21}\Gamma_{1,22} + g^{22}\Gamma_{2,22} = 0.$$

(12.63)

Substituting the results (12.63) into (12.59), we obtain

$$R_{1212} = -\frac{1}{2}\partial_1\partial_1 g_{22} + g_{22}(\Gamma_{12}^2)^2$$
$$R_{1212} = (\sin x^1)^2 - (\cos x^1)^2 + (\sin x^1)^2(\sin x^1)^{-2}(\cos x^1)^2$$
$$R_{1212} = (\sin x^1)^2.$$ (12.64)

Finally the Gauss curvature of the unit sphere is given by

$$\frac{R}{2} = \frac{1}{g}R_{1212} = (\sin x^1)^{-2}(\sin x^1)^2 = 1,$$ (12.65)

which proves that the Gauss curvature of the unit sphere is indeed equal to unity.

In the case of three-dimensional space ($N = 3$), in which the curvature tensor has a total of $3^3 = 27$ components, according to (12.51) there are six independent components of the covariant curvature tensor:

$$n = \frac{9(9-1)}{12} = 6. \qquad (12.66)$$

There are also six independent components of the Ricci tensor R_{mn}. Using the suitable Descartes coordinates in a given point, it is always possible to make three of these six components equal to zero. In particular, it is possible to diagonalize the Ricci tensor and define the curvature in any given point of a three-dimensional space by three independent quantities. In four-dimensional space there are 20 independent components of the covariant curvature tensor:

$$n = \frac{16(16-1)}{12} = 20. \qquad (12.67)$$

The case of a four-dimensional space is particularly important in the general theory of relativity and in cosmology.

Part III

Special Theory of Relativity

► Chapter 13

Relativistic Kinematics

13.1 | The Principle of Relativity

For the description of the motion of particles, it is necessary to have a system of reference. By a system of reference we mean a coordinate system to which we attach a clock. The coordinate system is used to determine the positions of particles in space and the clock is used to mesure the times at which the positions of particles in space are measured. There are systems of reference where the free motion of a particle, i.e., the motion of a particle on which there is no action of any external forces, is such that the particle has a constant velocity. Such systems of reference are called *inertial systems*.

If two systems of reference are translated with a constant velocity with respect to each other and one of them is an inertial system, then the other system of reference is also an inertial system. Any free motion in that system of reference will also be along a straight line with constant velocity.

Using these definitions, we can write down the statement of the *special principle of relativity* as follows:

All laws of nature are equal in all inertial systems of reference. In other words, the equations that express the laws of nature are invariant with respect to the transformations of spatial coordinates and time from one inertial system of reference to another.

Invariance of the Speed of Light

Let us consider two bodies interacting with each other, and let us assume that on one of the bodies some event has occurred (explosion, distance increase, or some other change). Then this event will be noticed at the position of the other body after the lapse of some time. If the distance between these two bodies is divided by this time interval, we obtain the *maximum speed of interaction.*

In nature, the motion of bodies with a speed higher than the maximum speed of interaction is not allowed, since by means of such a body it would be possible to achieve the interaction with a speed higher than the maximum speed of interaction. Based on the special principle of relativity, we conclude that the maximum speed of interaction is invariant with respect to the transformations of spatial coordinates and time from one inertial system of reference to another. Thus, it is a universal constant of nature. It is shown that this universal constant is equal to the *speed of light* in the vacuum, and its numerical value is given by

$$c = 2.99776 \times 10^8 \text{m/s}. \tag{13.1}$$

The *principle of invariance of the speed of light* is therefore stated as one of the basic principles of relativistic mechanics. According to this principle the speed of light is independent of the motion of the light source, i.e., of the choice of the system of reference in which the motion of light is described. In the next section, the mathematical formulation of this principle is presented.

13.3 The Interval between Events

An event in nature is determined by four coordinates. These include the three space coordinates of the position where the event has taken place and the time coordinate when the event has taken place. Each event is represented by a point, called the *world point* of the event, with the coordinates

$$x^m = \left\{ x^0, x^1, x^2, x^3 \right\} \quad (m = 0, 1, 2, 3). \tag{13.2}$$

The first coordinate is a time coordinate defined by $x^0 = ct$, and it has the dimension of length. The other three coordinates are the spatial Descartes coordinates

$$x^\alpha = \{x, y, z\} \quad (\alpha = 1, 2, 3). \tag{13.3}$$

We will from now on use Latin indices for the coordinates of the four-dimensional space-time (13.2), and Greek indices for the usual three-dimensional spatial coordinates (13.3).

The set of all world points constitutes a four-dimensional manifold called the *world*. To each particle in such a world, there corresponds a line called the *world line*. The world line of a particle at rest is a line parallel to the time axis.

In order to express the principle of invariance of the speed of light mathematically, let us now consider two systems of reference denoted by K and K', which move with respect to each other along the common x-axis with some constant velocity V. These coordinate systems are shown in Figure 4. The times in coordinate systems K and K' are denoted by t and t', respectively. Let one event consist of a signal sent at time t_1, with a constant velocity equal to the speed of light, from the point with coordinates $\{x_1, y_1, z_1\}$ in the coordinate system K. Let the other event consist of the same signal being received at time t_2 and at the point with coordinates $\{x_2, y_2, z_2\}$ in the coordinate system K.

In the coordinate system K, the coordinates of these two events are related by the following equation:

$$c^2(t_2 - t_1)^2 - (x_2 - x_1)^2 - (y_2 - y_1)^2 - (z_2 - z_1)^2 = 0. \qquad (13.4)$$

Because of the principle of invariance of the speed of light, in the coordinate system K' the coordinates of these two events are related by the

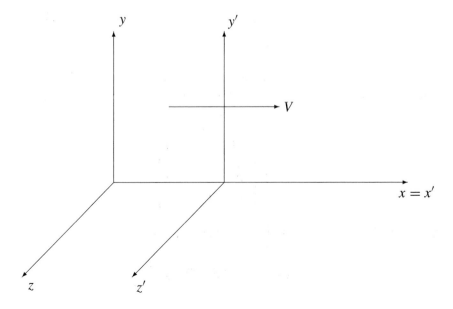

Figure 4. The coordinate systems K and K'

analogous equation

$$c^2(t_2' - t_1')^2 - (x_2' - x_1')^2 - (y_2' - y_1')^2 - (z_2' - z_1')^2 = 0. \quad (13.5)$$

If the coordinates (t_1, x_1, y_1, z_1) and (t_2, x_2, y_2, z_2) are the coordinates of any two events, then the quantity

$$s_{12} = \left[c^2(t_2 - t_1)^2 - (x_2 - x_1)^2 - (y_2 - y_1)^2 - (z_2 - z_1)^2 \right]^{1/2} \quad (13.6)$$

is called the *interval* between these two events. Analogously to (13.6), the square of the interval between two infinitesimally close events is determined by the metric

$$ds^2 = c^2 dt^2 - dx^2 - dy^2 - dz^2 = g_{mn} dx^m dx^n, \quad (13.7)$$

where $(m, n = 0, 1, 2, 3)$, and the contravariant space-time coordinates in this four-dimensional metric space are given by

$$x^m = \{x^0, x^1, x^2, x^3\} = \{ct, x, y, z\}. \quad (13.8)$$

The components of the covariant metric tensor g_{mn} are given by the following matrix:

$$[g_{mn}] = \begin{bmatrix} 1 & 0 & 0 & 0 \\ 0 & -1 & 0 & 0 \\ 0 & 0 & -1 & 0 \\ 0 & 0 & 0 & -1 \end{bmatrix}. \quad (13.9)$$

From (13.9) we conclude that the world is a four-dimensional pseudo-Euclidean metric space. The determinant of the metric tensor g_{mn} is $g = -1$. The matrix of cofactors of the matrix (13.9) is

$$[G^{mn}] = \begin{bmatrix} -1 & 0 & 0 & 0 \\ 0 & 1 & 0 & 0 \\ 0 & 0 & 1 & 0 \\ 0 & 0 & 0 & 1 \end{bmatrix}. \quad (13.10)$$

Using (6.13), the inverse matrix to the matrix $[g_{mn}]$ is given by

$$[g^{mn}] = \text{inv} \, [g_{mn}] = \frac{\text{adj} \, [g_{mn}]}{g} = \frac{[G^{mn}]^T}{g}. \quad (13.11)$$

We conclude that the components of the contravariant metric tensor g^{mn} are the same as the components of the covariant metric tensor (13.9), i.e.,

$$[g^{mn}] = \begin{bmatrix} 1 & 0 & 0 & 0 \\ 0 & -1 & 0 & 0 \\ 0 & 0 & -1 & 0 \\ 0 & 0 & 0 & -1 \end{bmatrix}. \tag{13.12}$$

From (13.9) and (13.12) we see that

$$g^{mk}g_{kn} = \delta_n^m \quad (k, m, n = 0, 1, 2, 3). \tag{13.13}$$

The covariant space-time coordinates in this four-dimensional metric space are given by

$$x_m = g_{mn}x^n = \{x_0, x_1, x_2, x_3\} = \{ct, -x, -y, -z\}. \tag{13.14}$$

Using (13.8) and (13.14) we see that

$$x_m x^m = g_{mn}x^m x^n = c^2 t^2 - x^2 - y^2 - z^2. \tag{13.15}$$

Using expressions (13.4) and (13.5) and the principle of invariance of the speed of light, we conclude that an interval that is equal to zero in one inertial system of reference is also equal to zero in any other inertial system of reference. On the other hand, the quantities ds and ds' are two infinitesimally small quantities of the same order. Thus we may write

$$ds = ads', \tag{13.16}$$

where the coefficient a may depend only on the absolute value of the relative velocity V of the two inertial systems of reference K and K'. It may not depend on the spatial coordinates or time because of the assumption of the homogeneity of space and time in the inertial systems of reference. It may not depend on the direction of the relative velocity V of the two inertial systems of reference K and K' either, because of the assumption of the isotropic nature of space and time in the inertial systems of reference. Thus, as we may write (13.16), we may also write

$$ds' = ads. \tag{13.17}$$

From the equations (13.16) and (13.17) we obtain $a^2 = 1$ or $a = \pm 1$. On the other hand, from the special case of the identity transformation with $ds = ds'$ we conclude that $a = +1$, so that we always have

$$ds' = ds. \tag{13.18}$$

Thus we obtain the mathematical formulation of the principle of invariance of the speed of light (13.18), which implies the invariance of the interval with respect to the transformations from one inertial system of reference to another.

13.4   **Lorentz Transformations**

The objective of the present section is to derive the formulae for transformations of coordinates from one inertial system of reference to another. In other words, if we know the coordinates of a certain event $x^m = \{ct', x', y', z'\}$ in some inertial system of reference K', we need the expressions for the coordinates of that event $z^m = \{ct, x, y, z\}$ in some other inertial system of reference K. As x^m is a contravariant vector, it is transformed according to the transformation law

$$z^m = \frac{\partial z^m}{\partial x^n} x^n = \Lambda_n^m x^n. \tag{13.19}$$

The transformation (13.19) is a linear transformation and the coefficients Λ_n^m are independent of coordinates. The system Λ_n^m is a mixed second-order tensor in the pseudo-Euclidean metric space defined by the metric (13.7), since it is defined with respect to the linear transformations (13.19).

In order to calculate the components of the tensor Λ_n^m, we now introduce an imaginary time coordinate

$$\tau = ict \quad (i = \sqrt{-1}), \tag{13.20}$$

and an imaginary metric

$$d\sigma = ids \quad (i = \sqrt{-1}). \tag{13.21}$$

Thus the metric of the space becomes

$$d\sigma^2 = d\tau^2 + dx^2 + dy^2 + dz^2, \tag{13.22}$$

and we thereby define the four-dimensional Descartes coordinates. The linear transformation (13.19) must keep the metric ds or $d\sigma$ invariant. As we have argued before, the only transformations of Descartes coordinates that keep the metric form invariant are the transformations of translation, rotation, and inversion.

Parallel translation of four-dimensional Descartes coordinates is not a suitable candidate for the transformation (13.19), since it merely changes the origin of the spatial coordinates and the origin for measurement of time. Similarly, the inversion of the four-dimensional Descartes coordinates is not suitable either, since it merely changes the sign of the spatial coordinates and time.

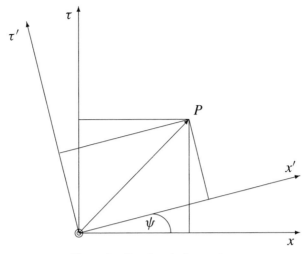

Figure 5.   Rotations in the τx plane

Thus we conclude that the only suitable candidate for the transformation
(13.19) is a rotation of four-dimensional Descartes coordinates. We are
looking for the formulae for transformation from the coordinate system
K' to the coordinate system K as shown in Figure 4. In that case we have
$y = y'$ and $z = z'$ and we are only interested for the rotation in the τx plane,
as shown in Figure 5.

From Figure 5, we immediately obtain the remaining two transformation
formulae:

$$x = x' \cos \psi - \tau' \sin \psi$$
$$\tau = x' \sin \psi + \tau' \cos \psi. \tag{13.23}$$

According to Figure 4, the coordinate system K' is moving with respect to
the coordinate system K with a constant velocity V. If we then consider
the motion of the origin of the coordinate system K' we then have $x' = 0$.
Then from (13.23) we obtain

$$\tan \psi = -\frac{x}{\tau} = i\frac{V}{c}. \tag{13.24}$$

Now, using the trigonometric formulae

$$\cos \psi = \frac{1}{\sqrt{1 + \tan^2 \psi}}, \quad \sin \psi = \frac{\tan \psi}{\sqrt{1 + \tan^2 \psi}}, \tag{13.25}$$

we obtain

$$\cos \psi = \frac{1}{\sqrt{1 - \frac{V^2}{c^2}}}, \quad \sin \psi = \frac{iV/c}{\sqrt{1 - \frac{V^2}{c^2}}}. \tag{13.26}$$

Substituting (13.26) into (13.23), we obtain

$$\begin{aligned}
\tau &= \frac{\tau' + i(V/c)x'}{\sqrt{1 - \frac{V^2}{c^2}}} & t &= \frac{t' + (V/c^2)x'}{\sqrt{1 - \frac{V^2}{c^2}}} \\
x &= \frac{x' - i(V/c)\tau'}{\sqrt{1 - \frac{V^2}{c^2}}} & x &= \frac{x' + Vt'}{\sqrt{1 - \frac{V^2}{c^2}}} \\
y &= y' & y &= y' \\
z &= z' & z &= z'
\end{aligned} \tag{13.27}$$

The results (13.27) are the well-known formulae for transformations of coordinates from one inertial system of reference to another. They are called the *Lorentz transformations*. In the special case when $V \ll c$, the Lorentz transformations are reduced into the so called *Gallilei transformations* of nonrelativistic mechanics

$$\begin{aligned}
t &= t' \\
x &= x' + Vt \\
y &= y' \\
z &= z'.
\end{aligned} \tag{13.28}$$

Although the Gallilei transformations are closer to our everyday experience than the Lorentz transformations, they are not in accordance with the principle of relativity and they do not leave the metric form of the four-dimensional space-time invariant.

Using (13.27), it is easy to construct the tensor of Lorentz transformations Λ_n^m in the following matrix form:

$$[\Lambda_n^m] = \begin{bmatrix} \frac{1}{\sqrt{1 - \frac{V^2}{c^2}}} & \frac{V/c}{\sqrt{1 - \frac{V^2}{c^2}}} & 0 & 0 \\ \frac{V/c}{\sqrt{1 - \frac{V^2}{c^2}}} & \frac{1}{\sqrt{1 - \frac{V^2}{c^2}}} & 0 & 0 \\ 0 & 0 & 1 & 0 \\ 0 & 0 & 0 & 1 \end{bmatrix}. \tag{13.29}$$

The inverse of the tensor Λ_n^m, denoted by $(\Lambda_n^m)^{-1}$, is given in the matrix form

$$\left[(\Lambda_n^m)^{-1}\right] = \begin{bmatrix} \dfrac{1}{\sqrt{1-\frac{v^2}{c^2}}} & -\dfrac{V/c}{\sqrt{1-\frac{v^2}{c^2}}} & 0 & 0 \\[3ex] -\dfrac{V/c}{\sqrt{1-\frac{v^2}{c^2}}} & \dfrac{1}{\sqrt{1-\frac{v^2}{c^2}}} & 0 & 0 \\[3ex] 0 & 0 & 1 & 0 \\[1ex] 0 & 0 & 0 & 1 \end{bmatrix}, \tag{13.30}$$

and it is obtained from the tensor (13.29) by reversing the sign of the relative velocity V, i.e., by putting $V \to -V$. Thus we may write

$$\frac{\partial x^m}{\partial z^n} = (\Lambda_n^m)^{-1}(V) = \Lambda_n^m(-V). \tag{13.31}$$

It is easily shown, by direct multiplication of the matrices (13.29) and (13.30), that

$$\Lambda_k^m (\Lambda_n^k)^{-1} = \delta_n^m. \tag{13.32}$$

13.5 Velocity and Acceleration Vectors

In the special theory of relativity, the time differential dt is not a scalar invariant, and the usual definition of the three-dimensional velocity

$$v^\alpha = \frac{dx^\alpha}{dt} \quad (\alpha = 1, 2, 3), \tag{13.33}$$

is less useful, since it does not behave as a vector with respect to the transformations from one inertial system of reference to another. Therefore we introduce a *four-velocity*, as a contravariant vector

$$u^m = \frac{Dx^m}{ds} = \frac{dx^m}{ds} \quad (m = 0, 1, 2, 3). \tag{13.34}$$

On the other hand, by definition, we have

$$ds^2 = g_{mn}dx^m dx^n = c^2 dt^2 + g_{\alpha\beta}dx^\alpha dx^\beta$$
$$= c^2 dt^2 - dx^2 - dy^2 - dz^2, \tag{13.35}$$

and the square of the intensity of the three-dimensional velocity is given by

$$v^2 = \frac{dx^2 + dy^2 + dz^2}{dt^2} = -g_{\alpha\beta}v^\alpha v^\beta. \tag{13.36}$$

From (13.35) with (13.36), we obtain

$$ds^2 = c^2 dt^2 \left(1 - \frac{v^2}{c^2}\right) = c^2 dt^2 \left(1 + \frac{g_{\alpha\beta} v^\alpha v^\beta}{c^2}\right). \tag{13.37}$$

Substituting (13.37) into (13.34), we obtain

$$u^m = \frac{dx^m}{cdt\sqrt{1 - \frac{v^2}{c^2}}} \quad (m = 0, 1, 2, 3). \tag{13.38}$$

The temporal zeroth component of the four-velocity is then given by

$$u^0 = \frac{1}{\sqrt{1 - \frac{v^2}{c^2}}}, \tag{13.39}$$

and the three spatial components of the four-velocity are given by

$$u^\alpha = \frac{v^\alpha}{c\sqrt{1 - \frac{v^2}{c^2}}}. \tag{13.40}$$

From (13.39) and (13.40) we conclude that the components of four-velocity are not independent of each other, but satisfy the equality

$$u_m u^m = g_{mn} u^m u^n = \frac{g_{mn} dx^m dx^n}{ds^2} = 1, \tag{13.41}$$

since we have

$$ds^2 = g_{mn} dx^m dx^n. \tag{13.42}$$

The *four-acceleration* of a particle is defined as a contravariant vector

$$w^m = \frac{Du^m}{ds} = \frac{du^m}{ds} \quad (m = 0, 1, 2, 3). \tag{13.43}$$

Here, using (13.38), we may write

$$w^m = \frac{1}{c\sqrt{1 - \frac{v^2}{c^2}}} \frac{du^m}{dt}, \tag{13.44}$$

or

$$w^m = \frac{1}{c^2\sqrt{1 - \frac{v^2}{c^2}}} \frac{d}{dt} \frac{1}{\sqrt{1 - \frac{v^2}{c^2}}} \frac{dx^m}{dt}. \tag{13.45}$$

The temporal zeroth component of the four-acceleration is then given by

$$w^0 = \frac{1}{c\sqrt{1 - \frac{v^2}{c^2}}} \frac{d}{dt} \frac{1}{\sqrt{1 - \frac{v^2}{c^2}}}, \tag{13.46}$$

and the three spatial components of the four-acceleration are given by

$$w^\alpha = \frac{1}{c^2\sqrt{1 - \frac{v^2}{c^2}}} \frac{d}{dt} \frac{v^\alpha}{\sqrt{1 - \frac{v^2}{c^2}}}. \tag{13.47}$$

From (13.41), we conclude that the vector of four-acceleration is always orthogonal to the vector of four-velocity. This can be shown by differentiating (13.41) with respect to the metric, i.e.,

$$\frac{d}{ds}\left(g_{mn}u^m u^n\right) = g_{mn}\frac{du^m}{ds}u^n + g_{mn}u^m\frac{du^n}{ds} = 0,$$

$$g_{mn}\frac{du^m}{ds}u^n + g_{nm}\frac{du^n}{ds}u^m = 2g_{mn}\frac{du^m}{ds}u^n = 0, \tag{13.48}$$

$$g_{mn}w^m u^n = w_n u^n = 0.$$

Thus the vectors w^n and u^n are indeed orthogonal to each other.

Chapter 14

Relativistic Dynamics

14.1 Lagrange Equations

Let us consider a free particle, i.e., a particle that is not under the influence of any forces. The equations describing the motion of this particle are obtained using the variational principle. The action integral is defined by (11.4), i.e.,

$$I = \int_{S_A}^{S_B} L\left(x^n, \frac{dx^n}{ds}\right) ds = \int_{S_A}^{S_B} L(x^n, u^n) ds, \qquad (14.1)$$

where u^n is the four-velocity of the particle. The equations of motion of the particle are obtained using the variational principle, i.e., the condition that the variation of the action integral is equal to zero, $\delta I = 0$. The Lagrange equations of motion of the particle are given by (11.12) in the form

$$\frac{\partial L}{\partial x^n} - \frac{d}{ds}\left(\frac{\partial L}{\partial u^n}\right) = 0 \quad (n = 0, 1, 2, 3). \qquad (14.2)$$

Let us now define the form of the Lagrangian function for a free particle. The Equation (14.2), derived using the variational principle applied to the action integral (14.1), are equal in all inertial systems of reference and the action integral is a scalar invariant with respect to the Lorentz transformations (13.29). Thus, the Lagrange function L itself is also a scalar invariant. Because of the homogeneity of the four-dimensional space-time in the inertial frames of reference, the Lagrange function cannot be a function of space-time coordinates x^n. Thus we may write

$$L = L(u^n) \quad (n = 0, 1, 2, 3). \qquad (14.3)$$

The only invariant that can be created using the four-velocity vector u^n is given by

$$u_n u^n = g_{kn} u^k u^n = 1 \quad (k, n = 0, 1, 2, 3). \tag{14.4}$$

Thus the action integral of a free particle is simply proportional to the arc length in the four-dimensional space-time manifold, i.e.,

$$I = -mc \int_{S_A}^{S_B} ds = -mc \int_{S_A}^{S_B} \sqrt{g_{kn} \frac{dx^k}{ds} \frac{dx^n}{ds}} ds, \tag{14.5}$$

where m is a mass parameter of the free particle. From (14.5) we see that the Lagrangian function is given by the expression

$$L = -mc\sqrt{g_{kn} u^k u^n}. \tag{14.6}$$

Using the Lagrangian (14.6) we may write

$$\frac{\partial L}{\partial u^n} = \frac{\partial}{\partial u^n} \left[-mc(g_{kj} u^k u^j)^{1/2} \right]$$

$$= -\frac{1}{2} mc(g_{kj} u^k u^j)^{-1/2} (g_{kn} u^k + g_{nj} u^j). \tag{14.7}$$

Here, using $g_{kj} u^k u^j = 1$ and the symmetry of the metric tensor $g_{nj} = g_{jn}$, we obtain

$$\frac{\partial L}{\partial u^n} = -\frac{1}{2} mc \, 2 g_{kn} u^k = -mc g_{kn} u^k = -mc u_n. \tag{14.8}$$

Thus we may write

$$-\frac{d}{ds} \frac{\partial L}{\partial u^n} = mc \frac{du_n}{ds}. \tag{14.9}$$

Using the Lagrangian (14.6) we may also write

$$\frac{\partial L}{\partial x^n} = 0. \tag{14.10}$$

Substituting (14.9) and (14.10) into the Lagrange equation (14.2), we obtain

$$mc \frac{du_n}{ds} = \frac{d}{ds} (mc u_n) = 0. \tag{14.11}$$

The *equations of motion of a free particle* are then given in the following form:

$$w^n = \frac{du^n}{ds} = 0 \quad (n = 0, 1, 2, 3). \tag{14.12}$$

From Equation (14.12) we see that a free particle moves along a straight line with constant velocity.

Let us now investigate the nonrelativistic limit of the action integral (14.5) when the particles are moving with low velocities ($v \ll c$). If we substitute (13.37) into (14.5) we obtain

$$I = -mc^2 \int_{t_A}^{t_B} \sqrt{1 - \frac{v^2}{c^2}} dt = \int_{t_A}^{t_B} L_T(x^\alpha, v^\alpha) dt \qquad (14.13)$$

where L_T is a Lagrangian defined with respect to the nonscalar time variable t, given by

$$L_T(x^\alpha, v^\alpha) = -mc^2 \sqrt{1 - \frac{v^2}{c^2}} = -mc^2 \sqrt{1 + \frac{g_{\alpha\beta} v^\alpha v^\beta}{c^2}}. \qquad (14.14)$$

In the nonrelativistic limit of particles moving with low velocities ($v \ll c$), we may use the approximation

$$\sqrt{1 - \frac{v^2}{c^2}} \approx 1 - \frac{v^2}{2c^2}. \qquad (14.15)$$

Substituting (14.15) into (14.13) we obtain

$$I = \int_{t_A}^{t_B} \left(-mc^2 + \frac{1}{2} mv^2 \right) dt. \qquad (14.16)$$

Thus the nonrelativistic approximation of the Lagrangian L_T can be written as follows:

$$L_T(x^\alpha, v^\alpha) \approx \frac{1}{2} mv^2 - mc^2. \qquad (14.17)$$

As the nonrelativistic Lagrangian is defined up to an arbitrary additive constant, the second term in (14.17) does not contribute to the nonrelativistic equations of motion and can be dropped in the nonrelativistic applications. Its physical significance in the special theory of relativity will be discussed later in this chapter. Thus the action integral (14.5) has the proper nonrelativistic limit.

14.2 **Energy–Momentum Vector**

14.2.1 Introduction and Definitions

In nonrelativistic mechanics there are a number of constants of motion. These include the energy and the momentum of the particle. The objective of the present section is to define the energy and the momentum of a particle in the framework of the special theory of relativity. Let us start with the action integral (14.13), in the form

$$I = \int_{t_A}^{t_B} L_T(x^\alpha, v^\alpha) dt. \tag{14.18}$$

The Lagrange equations of motion of a particle, analogous to Equation (11.12), defined in terms of the Lagrangian L_T with the nonscalar time variable t as the parameter, are given by

$$\frac{\partial L_T}{\partial x^\alpha} - \frac{d}{dt}\left(\frac{\partial L_T}{\partial v^\alpha}\right) = 0 \quad (\alpha = 1, 2, 3). \tag{14.19}$$

Let us now calculate the total time derivative of the Lagrangian L_T as follows:

$$\frac{dL_T}{dt} = \frac{\partial L_T}{\partial x^\alpha}\frac{dx^\alpha}{dt} + \frac{\partial L_T}{\partial v^\alpha}\frac{dv^\alpha}{dt}. \tag{14.20}$$

Using here the equations of motion (14.19) and the definition of the three-dimensional velocity (13.33), we obtain

$$\frac{dL_T}{dt} = \frac{d}{dt}\left(\frac{\partial L_T}{\partial v^\alpha}\right)v^\alpha + \frac{\partial L_T}{\partial v^\alpha}\frac{dv^\alpha}{dt} = \frac{d}{dt}\left(\frac{\partial L_T}{\partial v^\alpha}v^\alpha\right). \tag{14.21}$$

The result (14.21) can be rewritten in the form

$$\frac{d}{dt}\left(\frac{\partial L_T}{\partial v^\alpha}v^\alpha - L_T\right) = 0. \tag{14.22}$$

Thus the quantity

$$\mathcal{E} = \frac{\partial L_T}{\partial v^\alpha}v^\alpha - L_T = \text{Constant} \tag{14.23}$$

is a constant of motion. The quantity (14.23) can be recognized from non-relativistic mechanics as the *total energy* of the particle. Let us also define the *momentum* p^α, conjugate to the coordinate x^α, as follows:

$$\vec{p} = \frac{\partial L_T}{\partial \vec{v}} \Rightarrow p^\alpha = -g^{\alpha\beta}\frac{\partial L_T}{\partial v^\beta} \Rightarrow p_\alpha = -\frac{\partial L_T}{\partial v^\alpha} \tag{14.24}$$

where

$$\{p^\alpha\} = \{p_x, p_y, p_z\}, \quad \{p_\alpha\} = \{-p_x, -p_y, -p_z\}. \tag{14.25}$$

Substituting (14.24) into (14.23) we obtain

$$\mathcal{E} = -p_\alpha v^\alpha - L_T = -g_{\alpha\beta}p^\alpha v^\beta - L_T = \vec{p}\cdot\vec{v} - L_T. \tag{14.26}$$

Using the definition (14.14) of the Lagrangian L_T, we can calculate

$$p_\alpha = -\frac{\partial L_T}{\partial v^\alpha} = mc^2 \frac{\partial}{\partial v^\alpha} \left(1 + \frac{g_{\nu\beta} v^\nu v^\beta}{c^2}\right)^{1/2}, \tag{14.27}$$

or

$$p_\alpha = mc^2 \frac{1}{2} \left(1 + \frac{g_{\nu\beta} v^\nu v^\beta}{c^2}\right)^{-1/2} \frac{1}{c^2} g_{\nu\beta} \frac{\partial}{\partial v^\alpha} (v^\nu v^\beta). \tag{14.28}$$

Using the symmetry of the metric tensor $g_{\nu\beta} = g_{\beta\nu}$, we obtain

$$p_\alpha = m \frac{1}{2} \left(1 + \frac{g_{\nu\beta} v^\nu v^\beta}{c^2}\right)^{-1/2} 2 g_{\alpha\beta} v^\beta. \tag{14.29}$$

Using $v_\alpha = g_{\alpha\beta} v^\beta$, we finally obtain

$$p_\alpha = \frac{m v_\alpha}{\sqrt{1 - \frac{v^2}{c^2}}} \Rightarrow p^\alpha = \frac{m v^\alpha}{\sqrt{1 - \frac{v^2}{c^2}}}. \tag{14.30}$$

Substituting (14.30) and (14.14) into (14.26), we obtain the total energy of the particle as follows:

$$\begin{aligned}
\mathcal{E} &= -\frac{m v_\alpha v^\alpha}{\sqrt{1 - \frac{v^2}{c^2}}} + mc^2 \sqrt{1 - \frac{v^2}{c^2}} = \frac{m v^2}{\sqrt{1 - \frac{v^2}{c^2}}} + mc^2 \sqrt{1 - \frac{v^2}{c^2}} \\
&= \frac{mc^2}{\sqrt{1 - \frac{v^2}{c^2}}} - \frac{m\left(c^2 - v^2\right)}{\sqrt{1 - \frac{v^2}{c^2}}} + mc^2 \sqrt{1 - \frac{v^2}{c^2}} \\
&= \frac{mc^2}{\sqrt{1 - \frac{v^2}{c^2}}} - mc^2 \sqrt{1 - \frac{v^2}{c^2}} + mc^2 \sqrt{1 - \frac{v^2}{c^2}} = \frac{mc^2}{\sqrt{1 - \frac{v^2}{c^2}}}. \tag{14.31}
\end{aligned}$$

Thus the total energy of the particle is given by

$$\mathcal{E} = \frac{mc^2}{\sqrt{1 - \frac{v^2}{c^2}}}. \tag{14.32}$$

The particle at rest with $v = 0$ has the so-called rest energy $\mathcal{E}_0$ given by

$$\mathcal{E}_0 = mc^2. \tag{14.33}$$

The kinetic energy of the particle is obtained if the rest energy (14.33) is subtracted from the total energy (14.32). Thus we obtain

$$\mathcal{E}_K = \mathcal{E} - \mathcal{E}_0 = mc^2 \left(\frac{1}{\sqrt{1 - \frac{v^2}{c^2}}} - 1 \right). \tag{14.34}$$

For the particles moving with low velocities ($v \ll c$), we may write

$$\frac{1}{\sqrt{1 - \frac{v^2}{c^2}}} \approx 1 + \frac{v^2}{2c^2}. \tag{14.35}$$

Substituting (14.35) into (14.34) we obtain

$$\mathcal{E}_K = \tfrac{1}{2} m v^2. \tag{14.36}$$

In the same approximation ($v \ll c$), the components of the momentum of the particle are given by

$$p^\alpha = m v^\alpha. \tag{14.37}$$

The approximate results (14.36) and (14.37) are the well-known non-relativistic expressions for the kinetic energy and the momentum of a moving particle, respectively.

14.2.2 Transformations of Energy–Momentum

The results for the momentum of the free particle (14.30) and the energy of the free particle (14.32) were derived from the Lagrangian L_T, defined with respect to the nonscalar time variable t, in a noncovariant way. Now we want to define the energy and momentum as the constants of motion using the covariant Lagrangian L given by (14.6). In analogy with the definitions (14.24), we now define the *energy–momentum* four-vector p_n with the following covariant and contravariant components:

$$p_n = -\frac{\partial L}{\partial u^n} = mcu_n \Rightarrow p^n = -\frac{\partial L}{\partial u_n} = mcu^n \tag{14.38}$$

From Equation (14.11) we see that the four components of the energy–momentum of a free particle satisfy the equations

$$\frac{dp^n}{ds} = 0 \quad (n = 0, 1, 2, 3). \tag{14.39}$$

Thus for a free particle the four components of the energy–momentum four-vector are conserved. Using (13.39) here, we obtain the temporal zeroth component of the energy–momentum vector, which is proportional to the energy $\mathcal{E}$ of the particle, in the form

$$p^0 = mcu^0 = \frac{mc}{\sqrt{1 - \frac{v^2}{c^2}}} = \frac{\mathcal{E}}{c}, \tag{14.40}$$

where the energy of the particle is given by (14.32). Furthermore, using (13.40), we obtain the three spatial components of the energy–momentum four-vector, which are equal to the components of the three-dimensional momentum of the particle p^α given by (14.30), i.e.,

$$p^\alpha = mcu^\alpha = \frac{mv^\alpha}{\sqrt{1 - \frac{v^2}{c^2}}}. \tag{14.41}$$

From the results (14.40) and (14.41), we see that the energy and momentum of a particle are not two independent quantities as in nonrelativistic mechanics. In relativistic mechanics they are components of the same four-vector. It should be noted that the constant of motion analogous to the nonrelativistic result (14.23), is identically equal to zero:

$$\frac{\partial L}{\partial u^n} u^n - L = -p_n u^n - L \equiv 0. \tag{14.42}$$

The energy–momentum vector as a four-vector transforms with respect to the transformations from one inertial system of reference to another, according to the transformation law

$$p^k = \Lambda^k_n p'^n, \tag{14.43}$$

where p^k are the components of the energy–momentum tensor in the inertial system of reference K, while p'^k are the components of the energy–momentum tensor in the inertial system of reference K' moving along the common x-axis with a velocity V with respect to K. Here, using the explicit form of the tensor Λ^k_n given by (13.29), we obtain

$$\mathcal{E} = \frac{1}{\sqrt{1 - \frac{V^2}{c^2}}} \left(\mathcal{E}' + Vp'_x \right)$$

$$p_x = \frac{1}{\sqrt{1 - \frac{V^2}{c^2}}} \left(p'_x + \frac{V}{c^2} \mathcal{E}' \right) \tag{14.44}$$

$$p_y = p'_y$$

$$p_z = p'_z.$$

Using (14.38) with (13.41) we obtain

$$p^n p_n = m^2 c^2 u^n u_n = m^2 c^2. \tag{14.45}$$

From (14.45) we may write

$$\frac{\mathcal{E}^2}{c^2} + p^\alpha p_\alpha = m^2 c^2. \tag{14.46}$$

From the results (14.25) for the components of the three-dimensional momentum vector, we may write

$$p^\alpha p_\alpha = g_{\alpha\beta} p^\alpha p^\beta = -\vec{p} \cdot \vec{p} = -p^2. \tag{14.47}$$

Substituting (14.47) into (14.46), we obtain

$$\frac{\mathcal{E}^2}{c^2} - p^2 = m^2 c^2. \tag{14.48}$$

The relation between the energy and the three-dimensional momentum of a particle then becomes

$$\mathcal{E} = c\sqrt{p^2 + m^2 c^2}. \tag{14.49}$$

The function (14.49) is usually called the *Hamiltonian* of the particle and is denoted by H. Thus we may write

$$H = c\sqrt{p^2 + m^2 c^2} = \sqrt{p^2 c^2 + m^2 c^4}, \tag{14.50}$$

being the most usual definition of the Hamiltonian of a particle.

14.2.3 Conservation of Energy–Momentum

The conservation law of the momentum of a particle is a consequence of the homogeneity of space and the conservation law of the energy of a particle is a consequence of the homogeneity of time. Because of the homogeneity of space-time, the mechanical properties of a free particle remain unchanged after the translation of a particle from a point with coordinates x^n to a point with coordinates $x^n + \lambda^n$, where λ^n is an infinitesimally small constant four-vector. Thus the Lagrange function must be invariant with respect to this translation and its variation must be zero. Therefore we may write

$$\delta L = \frac{\partial L}{\partial x^k} \delta x^k = \frac{\partial L}{\partial x^k} \lambda^k = 0. \tag{14.51}$$

Here, using the equations of motion (14.2), we obtain

$$\delta L = \frac{d}{ds} \left[\frac{\partial L}{\partial u^k} \right] \lambda^k = 0. \tag{14.52}$$

Since λ^n is a nonzero infinitesimal four-vector, the expression (14.52) gives

$$-\frac{d}{ds}\left[\frac{\partial L}{\partial u^k}\right] = 0. \tag{14.53}$$

Using (14.38) we now obtain

$$-\frac{d}{ds}\frac{\partial L}{\partial u^n} = mc\frac{du_n}{ds} = \frac{dp_n}{ds} = 0. \tag{14.54}$$

The expression (14.54) then gives

$$\frac{dp^n}{ds} = \frac{1}{c\sqrt{1-\frac{v^2}{c^2}}}\frac{dp^n}{dt} = 0 \Rightarrow p^n = \text{Constant}. \tag{14.55}$$

Thus we obtain the three-dimensional momentum conservation law

$$\frac{d\vec{p}}{dt} = 0 \Rightarrow \vec{p} = \text{Constant}, \tag{14.56}$$

and the energy conservation law

$$\frac{d\mathcal{E}}{dt} = 0 \Rightarrow \mathcal{E} = \text{Constant}. \tag{14.57}$$

14.3 Angular Momentum Tensor

In relativistic mechanics the *angular momentum tensor* is defined by the expression

$$M_{nk} = x_n p_k - x_k p_n \quad (k, n = 0, 1, 2, 3). \tag{14.58}$$

Only the spatial components of the angular momentum tensor with $(k, n = 1, 2, 3)$ have a physical meaning and coincide with the usual definition of the angular momentum in nonrelativistic mechanics. In nonrelativistic mechanics it is customary to form an axial three-dimensional angular momentum vector

$$M^\nu = \tfrac{1}{2}e^{\nu\alpha\beta}M_{\alpha\beta} = e^{\nu\alpha\beta}x_\alpha p_\beta \quad (\alpha, \beta = 1, 2, 3), \tag{14.59}$$

or, in the classical vector notation,

$$\vec{M} = \begin{vmatrix} \vec{1} & \vec{2} & \vec{3} \\ x_1 & x_2 & x_3 \\ p_1 & p_2 & p_3 \end{vmatrix} = \vec{r} \times \vec{p}. \tag{14.60}$$

The conservation law of the angular momentum tensor (14.58) is a consequence of the isotropic nature of the four-dimensional space-time.

The three-dimensional angular momentum conservation law is a consequence of the isotropic nature of the three-dimensional space. Because of the isotropic nature of the four-dimensional space-time, the mechanical properties of a free particle remain unchanged after rotations in the four-dimensional space-time. Let us consider the special case of a rotation for some angle θ about the 3-axis in three-dimensional space. Then the relation between the coordinates z^m in the rotated inertial system of reference K' and the coordinates x^k in the original inertial system of reference K is given by

$$z^n = \Omega^n_j x^j \quad (j, n = 0, 1, 2, 3). \tag{14.61}$$

Using (5.40) we may rewrite (14.61) in the matrix form

$$
\begin{bmatrix} z^0 \\ z^1 \\ z^2 \\ z^3 \end{bmatrix} =
\begin{bmatrix}
1 & 0 & 0 & 0 \\
0 & \cos\theta & \sin\theta & 0 \\
0 & -\sin\theta & \cos\theta & 0 \\
0 & 0 & 0 & 1
\end{bmatrix}
\begin{bmatrix} x^0 \\ x^1 \\ x^2 \\ x^3 \end{bmatrix}. \tag{14.62}
$$

If we now consider the rotation for some infinitesimal angle $\delta\theta \approx 0$, then we have

$$\cos\delta\theta \approx 1, \quad \sin\delta\theta \approx \delta\theta. \tag{14.63}$$

Substituting (14.63) into (14.62) we obtain

$$
\begin{bmatrix} z^0 \\ z^1 \\ z^2 \\ z^3 \end{bmatrix} =
\begin{bmatrix}
1 & 0 & 0 & 0 \\
0 & 1 & \delta\theta & 0 \\
0 & -\delta\theta & 1 & 0 \\
0 & 0 & 0 & 1
\end{bmatrix}
\begin{bmatrix} x^0 \\ x^1 \\ x^2 \\ x^3 \end{bmatrix}, \tag{14.64}
$$

or

$$z^n = x^n + \delta\Omega^n_j x^j = x^n + \delta\Omega^{nk} g_{kj} x^j = x^n + \delta\Omega^{nk} x_k \tag{14.65}$$

where $\delta\Omega^n_j$ is a mixed tensor defined by the following antisymmetric matrix:

$$
\left[\delta\Omega^n_j\right] =
\begin{bmatrix}
0 & 0 & 0 & 0 \\
0 & 0 & \delta\theta & 0 \\
0 & -\delta\theta & 0 & 0 \\
0 & 0 & 0 & 0
\end{bmatrix}. \tag{14.66}
$$

Here, using (13.9) and (13.12), we obtain the covariant coordinates x_k of the four-dimensional space-time in matrix form:

$$[x_k] = [g_{jk}]\left[x^j\right] = \begin{bmatrix} 1 & 0 & 0 & 0 \\ 0 & -1 & 0 & 0 \\ 0 & 0 & -1 & 0 \\ 0 & 0 & 0 & -1 \end{bmatrix} \begin{bmatrix} x^0 \\ x^1 \\ x^2 \\ x^3 \end{bmatrix} = \begin{bmatrix} x^0 \\ -x^1 \\ -x^2 \\ -x^3 \end{bmatrix}. \tag{14.67}$$

We can also calculate the components of the antisymmetric contravariant tensor $\delta\Omega^{nk}$ in matrix form as follows:

$$\left[\delta\Omega^{nk}\right] = \left[g^{jk}\right]\left[\delta\Omega_j^n\right], \tag{14.68}$$

or

$$\left[\delta\Omega^{nk}\right] = \begin{bmatrix} 1 & 0 & 0 & 0 \\ 0 & -1 & 0 & 0 \\ 0 & 0 & -1 & 0 \\ 0 & 0 & 0 & -1 \end{bmatrix} \begin{bmatrix} 0 & 0 & 0 & 0 \\ 0 & 0 & \delta\theta & 0 \\ 0 & -\delta\theta & 0 & 0 \\ 0 & 0 & 0 & 0 \end{bmatrix}. \tag{14.69}$$

Thus we finally obtain the matrix form of the antisymmetric contravariant tensor $\delta\Omega^{nk}$ as follows:

$$\left[\delta\Omega^{nk}\right] = \begin{bmatrix} 0 & 0 & 0 & 0 \\ 0 & 0 & -\delta\theta & 0 \\ 0 & \delta\theta & 0 & 0 \\ 0 & 0 & 0 & 0 \end{bmatrix}. \tag{14.70}$$

For an arbitrary infinitesimal rotation, the contravariant tensor $\delta\Omega^{nk}$ has a more complex form compared to the simple matrix (14.70), but it is always an antisymmetric tensor. Thus we always have

$$\delta\Omega^{nk} = -\delta\Omega^{kn} \quad (k, n = 0, 1, 2, 3), \tag{14.71}$$

and the most general infinitesimal rotation of the coordinates is given by the transformation relations

$$z^n = x^n + \delta x^n, \quad \delta x^n = \delta\Omega^{nk} x_k. \tag{14.72}$$

The most general infinitesimal rotation of four-velocity vector is given by the following transformation relations:

$$u'^n = u^n + \delta u^n, \quad \delta u^n = \delta\Omega^{nk} u_k. \tag{14.73}$$

The Lagrange function of a particle $L(x^n, u^n)$ must be invariant with respect to this infinitesimal rotation, i.e., we must have $\delta L = 0$. Let us now calculate the variation of the Lagrangian L with respect to the infinitesimal rotations (14.72) and (14.73), i.e.,

$$\delta L = \frac{\partial L}{\partial x^n} \delta x^n + \frac{\partial L}{\partial u^n} \delta u^n = 0, \qquad (14.74)$$

or

$$\delta L = \frac{\partial L}{\partial x^n} \delta \Omega^{nk} x_k + \frac{\partial L}{\partial u^n} \delta \Omega^{nk} u_k = 0. \qquad (14.75)$$

Substituting the equations of motion (14.2) into (14.75) we obtain

$$\delta L = \delta \Omega^{nk} \left[\frac{d}{ds} \left(\frac{\partial L}{\partial u^n} \right) x_k + \frac{\partial L}{\partial u^n} u_k \right] = 0. \qquad (14.76)$$

Using the definition of the energy–momentum four-vector (14.38) and the definition of the four-velocity (13.34), we obtain

$$\delta L = -\delta \Omega^{nk} \left[\frac{dp_n}{ds} x_k + p_n \frac{dx_k}{ds} \right] = -\frac{d}{ds} \left(\delta \Omega^{nk} p_n x_k \right) = 0. \qquad (14.77)$$

Using here the antisymmetry of the tensor $\delta \Omega^{nk}$, the Equation (14.77) becomes

$$\delta L = -\frac{1}{2} \delta \Omega^{nk} \frac{d}{ds} (p_n x_k - p_k x_n) = \frac{1}{2} \delta \Omega^{nk} \frac{dM_{nk}}{ds} = 0. \qquad (14.78)$$

From (14.78), as a direct consequence of the isotropic nature of space-time, we obtain the conservation law of the angular momentum tensor, i.e.,

$$\frac{dM_{nk}}{ds} = 0 \Rightarrow \frac{dM_{nk}}{dt} = 0 \Rightarrow M_{nk} = \text{Constant}. \qquad (14.79)$$

The spatial part of this conservation law for $(n, k = 1, 2, 3)$, i.e.,

$$M^\nu = e^{\nu\alpha\beta} x_\alpha p_\beta = \text{Constant}, \qquad (14.80)$$

is the usual conservation law of the three-dimensional angular momentum vector.

Chapter 15

Electromagnetic Fields

15.1   **Electromagnetic Field Tensor**

The electromagnetic field in the four-dimensional space-time is described by a four-vector

$$A^n = A^n(x^k), \qquad (15.1)$$

which is called the *potential of the electromagnetic field*. The temporal component of this four-vector is defined by

$$A^0 = \frac{\psi(x^k)}{c}, \qquad (15.2)$$

where $\psi(x^k)$ is the *electric scalar potential*. The three spatial components of this four-vector,

$$A^\alpha = A^\alpha(x^k), \qquad (15.3)$$

constitute the so-called *magnetic vector potential*. Thus we may write

$$A^n = \left\{ \frac{1}{c}\psi, A_x, A_y, A_z \right\}$$

$$A_n = g_{nk}A^k = \left\{ \frac{1}{c}\psi, -A_x, -A_y, -A_z \right\}. \qquad (15.4)$$

The action for a particle with mass m and charge q moving in the electromagnetic field defined by $A^n(x^k)$ is then given by

$$I = I_S + I_Q, \tag{15.5}$$

where I_S is the action for a free particle given by

$$I_S = -mc \int_{S_A}^{S_B} \sqrt{g_{kn}u^k u^n} \, ds, \tag{15.6}$$

and I_Q is the action describing the interaction of the charged particle with the electromagnetic field defined by the four-vector $A^n(x^k)$. The simplest invariant action that can be constructed using four-vectors $A^n(x^k)$ and u^n, describing the electromagnetic field and the motion of the particle, respectively, is given by

$$I_Q = -q \int_{S_A}^{S_B} g_{kn} A^k u^n \, ds = -q \int_{S_A}^{S_B} A_n u^n \, ds. \tag{15.7}$$

Substituting (15.6) and (15.7) into (15.5) we obtain

$$I = - \int_{S_A}^{S_B} \left(mc\sqrt{g_{kn}u^k u^n} + qA_n u^n \right) ds. \tag{15.8}$$

The Lagrangian of the charged particle moving in the electromagnetic field defined by $A^n(x^k)$ is then given by

$$L = -mc\sqrt{g_{kn}u^k u^n} - qA_n u^n. \tag{15.9}$$

Using (15.9) we obtain

$$\frac{\partial L}{\partial x^k} = -q\frac{\partial A_n}{\partial x^k}u^n, \tag{15.10}$$

and

$$\frac{d}{ds}\left(\frac{\partial L}{\partial u^k}\right) = -\frac{d}{ds}(mcu_k + qA_k). \tag{15.11}$$

Substituting (15.10) and (15.11) into the Lagrange Equations (14.2) gives

$$\frac{d}{ds}(mcu_k + qA_k) = q\frac{\partial A_n}{\partial x^k}u^n, \tag{15.12}$$

or

$$mc\frac{du_k}{ds} = q\left(\frac{\partial A_n}{\partial x^k} - \frac{\partial A_k}{\partial x^n}\right)u^n. \tag{15.13}$$

In (15.13) we define the covariant *electromagnetic field tensor* F_{kn} as follows:

$$F_{kn} = \frac{\partial A_n}{\partial x^k} - \frac{\partial A_k}{\partial x^n} = \partial_k A_n - \partial_n A_k. \qquad (15.14)$$

By definition (15.14), the covariant electromagnetic field tensor F_{kn} is an antisymmetric tensor and it satisfies

$$F_{kn} = -F_{nk} \quad (k, n = 0, 1, 2, 3). \qquad (15.15)$$

Substituting (15.14) into (15.13) we obtain the equations of motion of a charged particle in the electromagnetic field

$$mc\frac{du_k}{ds} = qF_{kn}u^n. \qquad (15.16)$$

Using the definition of four-momentum $p_k = mcu_k$ in (15.16), we may write

$$\frac{dp_k}{ds} = qF_{kn}u^n. \qquad (15.17)$$

Let us now demonstrate that the three spatial Equations (15.17) in the three-dimensional vector notation are equal to the well-known expression for the Lorentz force. Equation (15.17) for the three spatial components ($\alpha = 1, 2, 3$) has the form

$$\frac{dp_\alpha}{ds} = qF_{\alpha n}u^n = qF_{\alpha 0}u^0 + qF_{\alpha\beta}u^\beta. \qquad (15.18)$$

Using here (13.37), (13.39), and (13.40), we obtain

$$\frac{dp_\alpha}{dt} = qcF_{\alpha 0} + qF_{\alpha\beta}v^\beta. \qquad (15.19)$$

We then define the three-dimensional *electric field vector* denoted by E_α and the three-dimensional *magnetic induction vector* denoted by B^ν as follows:

$$F_{\alpha 0} = \frac{E_\alpha}{c}, \quad F^{\alpha 0} = \frac{E^\alpha}{c}$$

$$F_{\alpha\beta} = -e_{\alpha\beta\nu}B^\nu, \quad F^{\alpha\beta} = -e^{\alpha\beta\nu}B_\nu, \qquad (15.20)$$

where

$$\{E^\alpha\} = \{E_x, E_y, E_z\}, \quad \{E_\alpha\} = \{-E_x, -E_y, -E_z\}$$
$$\{B^\alpha\} = \{B_x, B_y, B_z\}, \quad \{B_\alpha\} = \{-B_x, -B_y, -B_z\}. \qquad (15.21)$$

Substituting (15.20) into (15.19) we obtain

$$\frac{dp_\alpha}{dt} = qE_\alpha - qe_{\alpha\beta\nu}v^\beta B^\nu. \tag{15.22}$$

Using here $p_\alpha = \{-p_x, -p_y, -p_z\}$, $v^\alpha = \{v_x, v_y, v_z\}$ and the expressions (15.21) for the components of the vectors $\vec{E}$ and $\vec{B}$, Equation (15.22) gives

$$-\frac{dp_x}{dt} = -qE_x - q\,(v_y B_z - v_z B_y)$$

$$-\frac{dp_y}{dt} = -qE_y - q\,(v_z B_x - v_x B_z) \tag{15.23}$$

$$-\frac{dp_z}{dt} = -qE_z - q\,(v_x B_y - v_y B_x).$$

Thus we obtain the familiar expression for the Lorentz force in the three-dimensional vector notation, as follows:

$$\frac{d\vec{p}}{dt} = q\vec{E} + q\vec{v} \times \vec{B}. \tag{15.24}$$

Using the definitions (15.20), the covariant electromagnetic field tensor F_{kn} can be written in the following matrix form:

$$[F_{kn}] = \begin{bmatrix} 0 & E_x/c & E_y/c & E_z/c \\ -E_x/c & 0 & -B_z & B_y \\ -E_y/c & B_z & 0 & -B_x \\ -E_z/c & -B_y & B_x & 0 \end{bmatrix}. \tag{15.25}$$

The mixed electromagnetic field tensor F_n^k is given by

$$F_n^k = g^{kj} F_{jn} \quad (j, k, n = 0, 1, 2, 3). \tag{15.26}$$

and can be written in the following matrix form:

$$[F_n^k] = [g^{kj}][F_{jn}] = \begin{bmatrix} 0 & E_x/c & E_y/c & E_z/c \\ E_x/c & 0 & B_z & -B_y \\ E_y/c & -B_z & 0 & B_x \\ E_z/c & B_y & -B_x & 0 \end{bmatrix}. \tag{15.27}$$

The contravariant electromagnetic field tensor F^{kn} is given by

$$F^{kn} = g^{kj} F_{jl} g^{ln} = F_l^k g^{ln} \quad (j, k, l, n = 0, 1, 2, 3), \tag{15.28}$$

and it can be written in the following matrix form:

$$[F^{kn}] = [F_l^k][g^{ln}] = \begin{bmatrix} 0 & -E_x/c & -E_y/c & -E_z/c \\ E_x/c & 0 & -B_z & B_y \\ E_y/c & B_z & 0 & -B_x \\ E_z/c & -B_y & B_x & 0 \end{bmatrix}. \qquad (15.29)$$

As the next step, let us now calculate the electric field vector $\vec{E}$ in terms of the potentials of the electromagnetic field (15.1). By definition (15.14) of the tensor F_{kn}, we have

$$E_\alpha = cF_{\alpha 0} = c\left(\frac{\partial A_0}{\partial x^\alpha} - \frac{\partial A_\alpha}{\partial x^0}\right). \qquad (15.30)$$

Using now $\psi = cA_0$ and $x^0 = ct$ we obtain

$$E_\alpha = \frac{\partial \psi}{\partial x^\alpha} - \frac{\partial A_\alpha}{\partial t}, \qquad (15.31)$$

or

$$E^\alpha = g^{\alpha\beta}\frac{\partial \psi}{\partial x^\beta} - \frac{\partial A^\alpha}{\partial t} = (\text{grad }\psi)^\alpha - \frac{\partial A^\alpha}{\partial t}. \qquad (15.32)$$

The contravariant components of the four-dimensional gradient of a scalar function ψ are given by

$$g^{kn}\frac{\partial \psi}{\partial x^n} = \left\{\frac{1}{c}\frac{\partial \psi}{\partial t}, -\text{grad}_x\,\psi, -\text{grad}_y\,\psi, -\text{grad}_z\,\psi\right\}. \qquad (15.33)$$

Using (15.33) for ($\alpha = 1, 2, 3$), we obtain the relation between the electric field vector $\vec{E}$ and the potentials of the electromagnetic field (15.1), in the three-dimensional vector notation

$$\vec{E} = -\frac{\partial \vec{A}}{\partial t} - \text{grad }\psi. \qquad (15.34)$$

Let us now calculate the magnetic induction vector $\vec{B}$ in terms of the potentials of the electromagnetic field (15.1). By definition (15.14) of the tensor F_{kn}, we have

$$F_{\alpha\beta} = \partial_\alpha A_\beta - \partial_\beta A_\alpha = \left(\delta_\alpha^\sigma \delta_\beta^\tau - \delta_\beta^\sigma \delta_\alpha^\tau\right)\partial_\sigma A_\tau$$
$$= e_{\nu\alpha\beta}e^{\nu\sigma\tau}\partial_\sigma A_\tau = e_{\alpha\beta\nu}e^{\nu\sigma\tau}\partial_\sigma A_\tau = -e_{\alpha\beta\nu}B^\nu \qquad (15.35)$$

where we have used the identity (10.45) in the form

$$e_{\nu\alpha\beta}e^{\nu\sigma\tau} = \delta_\alpha^\sigma \delta_\beta^\tau - \delta_\beta^\sigma \delta_\alpha^\tau. \qquad (15.36)$$

From (15.35) we obtain

$$B^\nu = -e^{\nu\sigma\tau}\partial_\sigma A_\tau. \tag{15.37}$$

Using (15.37) and the definition of the curl of the vector $\vec{A}$, we may write

$$B_x = -\left[\frac{\partial(-A_z)}{\partial y} - \frac{\partial(-A_y)}{\partial z}\right] = \frac{\partial A_z}{\partial y} - \frac{\partial A_y}{\partial z} = \text{curl}_x\, \vec{A}$$

$$B_y = -\left[\frac{\partial(-A_x)}{\partial z} - \frac{\partial(-A_z)}{\partial x}\right] = \frac{\partial A_x}{\partial z} - \frac{\partial A_z}{\partial x} = \text{curl}_y\, \vec{A}$$

$$B_z = -\left[\frac{\partial(-A_y)}{\partial x} - \frac{\partial(-A_x)}{\partial y}\right] = \frac{\partial A_y}{\partial x} - \frac{\partial A_x}{\partial y} = \text{curl}_z\, \vec{A}. \tag{15.38}$$

Equations (15.38), written in the three-dimensional vector notation give

$$\vec{B} = \text{curl}\, \vec{A}. \tag{15.39}$$

The result (15.39) is the familiar relation between the magnetic induction vector $\vec{B}$ and the magnetic vector potential of the electromagnetic field (15.3).

The temporal zeroth component of Equation (15.16) gives

$$mc^2\frac{du_0}{ds} = qcF_{0\alpha}u^\alpha, \tag{15.40}$$

or

$$\frac{d}{dt}\left(\frac{mc^2}{\sqrt{1-\frac{v^2}{c^2}}}\right) = \frac{d}{dt}\left(\frac{mc^2}{\sqrt{1-\frac{v^2}{c^2}}} - mc^2\right) = qcF_{0\alpha}v^\alpha. \tag{15.41}$$

From (15.20) we see that $cF_{0\alpha}v^\alpha = \vec{E}\cdot\vec{v}$. Thus we finally obtain

$$\frac{d\mathcal{E}_K}{dt} = q\vec{E}\cdot\vec{v}. \tag{15.42}$$

Equation (15.42) is the statement that the change of the kinetic energy $\mathcal{E}_K$ of a charged particle in the electromagnetic field is equal to the work done by the electric field $\vec{E}$. The work done by the magnetic field is identically equal to zero, since the force of the magnetic field $q\vec{v}\times\vec{B}$ is always perpendicular to the direction of the velocity $\vec{v}$.

15.2   Gauge Invariance

From the result (15.24) we see that, by measurements of the forces acting on a charged particle in the electromagnetic field, we can measure the components of the three-dimensional electric field vector $\vec{E}$ and the three-dimensional magnetic induction vector $\vec{B}$. Thus the components of the electromagnetic field tensor are observable physical quantities that are uniquely defined. On the other hand, the components of the potential of the electromagnetic field (15.1) are not uniquely defined. From the definition of the electromagnetic field tensor (15.14) we see that its components are invariant with respect to the *gauge transformations* of the potential of the electromagnetic field:

$$\bar{A}_n = A_n + \frac{\partial \phi}{\partial x^n}, \tag{15.43}$$

where ϕ is an arbitrary scalar function. Substituting (15.43) into (15.14), we obtain

$$\bar{F}_{kn} = \frac{\partial \bar{A}_n}{\partial x^k} - \frac{\partial \bar{A}_k}{\partial x^n} = \frac{\partial A_n}{\partial x^k} - \frac{\partial A_k}{\partial x^n} = F_{kn}. \tag{15.44}$$

Thus we need to impose an additional condition on the components of the potential of the electromagnetic field in order to make it more precisely defined. Such a condition is usually called the *gauge condition* or simply the *gauge* of the theory. In relativistic electrodynamics the most commonly used gauge is the *Lorentz gauge*, which requires that the components of the potential of the electromagnetic field satisfy the equation

$$\frac{\partial A_n}{\partial x_n} = \frac{\partial A^n}{\partial x^n} = \frac{\partial A^0}{\partial x^0} + \frac{\partial A^\alpha}{\partial x^\alpha} = \operatorname{div} \vec{A} + \frac{1}{c^2} \frac{\partial \psi}{\partial t} = 0. \tag{15.45}$$

When the Lorentz gauge (15.45) is adopted, then using (15.43) we obtain

$$\frac{\partial \bar{A}_n}{\partial x_n} = \frac{\partial A_n}{\partial x_n} + \frac{\partial^2 \phi}{\partial x_n \partial x^n}. \tag{15.46}$$

As both potentials $\bar{A}(x^k)$ and $A(x^k)$ satisfy the Lorentz condition (15.45), the arbitrary function $\phi(x^k)$ must satisfy the wave equation of the form

$$-\frac{\partial^2 \phi}{\partial x_n \partial x^n} = \nabla^2 \phi - \frac{1}{c^2} \frac{\partial^2 \phi}{\partial t^2} = 0. \tag{15.47}$$

Thus $\phi(x^k)$ is no longer an arbitrary scalar function but a solution to the wave Equation (15.47). It should be noted here that, even when we impose the Lorentz condition (15.45), the components of the potential of the electromagnetic field are still not uniquely defined and only the class of the allowed gauge transformations (15.43) is significantly reduced.

15.3 | Lorentz Transformations and Invariants

The potential of the electromagnetic field A^k, as a four-vector, transforms with respect to the transformations from one inertial system of reference to another according to the transformation law

$$A^k = \Lambda^k_n A'^n \quad (k, n = 0, 1, 2, 3), \tag{15.48}$$

where A^k are the components of the energy–momentum tensor in the inertial system of reference K, while A'^n are the components of the energy–momentum tensor in the inertial system of reference K' moving along the common x-axis with a velocity V with respect to K. Using here the explicit form of the tensor Λ^k_n given by (13.29), we obtain

$$\psi = \frac{1}{\sqrt{1 - \frac{V^2}{c^2}}} \left(\psi' + V A'_x \right)$$

$$A_x = \frac{1}{\sqrt{1 - \frac{V^2}{c^2}}} \left(A'_x + \frac{V}{c^2} \psi' \right) \tag{15.49}$$

$$A_y = A'_y$$

$$A_z = A'_z.$$

The transverse components of the vector $\vec{A}$ remain unchanged, and only the longitudinal component is affected by the transformation from the inertial system of reference K to the inertial system of reference K', moving along the common x-axis with a velocity V with respect to K.

The electromagnetic field tensor F_{kn} transforms from one inertial system of reference to another, according to the law

$$F_{jl} = \Lambda^k_j \Lambda^n_l F'_{kn} \quad (j, k, l, n = 0, 1, 2, 3), \tag{15.50}$$

where F_{jl} are the components of the energy–momentum tensor in the inertial system of reference K while F'_{kn} are the components of the energy–momentum tensor in the inertial system of reference K' moving along the common x-axis with a velocity V with respect to K. Using here the explicit form of the tensor Λ^k_n given by (13.29), we obtain

$$
\begin{aligned}
F_{01} &= \Lambda^k_0 \Lambda^n_1 F'_{kn} = \Lambda^0_0 \Lambda^1_1 F'_{01} + \Lambda^1_0 \Lambda^0_1 F'_{10} \\
F_{02} &= \Lambda^k_0 \Lambda^n_2 F'_{kn} = \Lambda^0_0 \Lambda^2_2 F'_{02} + \Lambda^1_0 \Lambda^2_2 F'_{12} \\
F_{03} &= \Lambda^k_0 \Lambda^n_3 F'_{kn} = \Lambda^0_0 \Lambda^3_3 F'_{03} + \Lambda^1_0 \Lambda^3_3 F'_{13} \\
F_{23} &= \Lambda^k_2 \Lambda^n_3 F'_{kn} = \Lambda^2_2 \Lambda^3_3 F'_{23} \\
F_{31} &= \Lambda^k_3 \Lambda^n_1 F'_{kn} = \Lambda^3_3 \Lambda^1_1 F'_{31} + \Lambda^3_3 \Lambda^0_1 F'_{30} \\
F_{12} &= \Lambda^k_1 \Lambda^n_2 F'_{kn} = \Lambda^1_1 \Lambda^2_2 F'_{12} + \Lambda^0_1 \Lambda^2_2 F'_{02}.
\end{aligned}
\tag{15.51}
$$

Introducing here the notation

$$\gamma = \frac{1}{\sqrt{1 - \frac{V^2}{c^2}}} \Rightarrow \gamma^2 \left(1 - \frac{V^2}{c^2}\right) = 1, \tag{15.52}$$

we may write

$$\Lambda_0^0 = \Lambda_1^1 = \gamma, \quad \Lambda_0^1 = \Lambda_1^0 = \frac{V}{c}\gamma, \quad \Lambda_2^2 = \Lambda_3^3 = 1. \tag{15.53}$$

Substituting (15.53) into (15.51), we obtain

$$F_{01} = \gamma^2 \left(1 - \frac{V^2}{c^2}\right) F_{01}' = F_{01}'$$

$$F_{02} = \gamma F_{02}' + \frac{V}{c}\gamma F_{12}'$$

$$F_{03} = \gamma F_{03}' + \frac{V}{c}\gamma F_{13}' \tag{15.54}$$

$$F_{23} = F_{23}'$$

$$F_{31} = \gamma F_{31}' + \frac{V}{c}\gamma F_{30}'$$

$$F_{12} = \gamma F_{12}' + \frac{V}{c}\gamma F_{02}'.$$

Now, using the explicit form of the tensor F_{kn} given by (15.25), we have

$$E_x = E_x'$$

$$E_y = \gamma \left(E_y' - VB_z'\right)$$

$$E_z = \gamma \left(E_z' + VB_y'\right)$$

$$B_x = B_x' \tag{15.55}$$

$$B_y = \gamma \left(B_y' + \frac{V}{c^2}E_z'\right)$$

$$B_z = \gamma \left(B_z' - \frac{V}{c^2}E_y'\right).$$

Thus, we finally obtain the transformation laws for the three-dimensional electric field vector $\vec{E}$ and the three-dimensional magnetic induction vector $\vec{B}$ in the form

$$E_x = E_x', \quad E_y = \frac{E_y' - VB_z'}{\sqrt{1 - \frac{V^2}{c^2}}}, \quad E_z = \frac{E_z' + VB_y'}{\sqrt{1 - \frac{V^2}{c^2}}} \tag{15.56}$$

$$B_x = B_x', \quad B_y = \frac{B_y' + \frac{V}{c^2}E_z'}{\sqrt{1 - \frac{V^2}{c^2}}}, \quad B_z = \frac{B_z' - \frac{V}{c^2}E_y'}{\sqrt{1 - \frac{V^2}{c^2}}}. \tag{15.57}$$

The longitudinal components of the vectors $\vec{E}$ and $\vec{B}$ remain unchanged, and only the transverse components are affected by the transformation from the inertial system of reference K to the inertial system of reference K', moving along the common x-axis with a velocity V with respect to K.

In order to find the invariants of the electromagnetic field, following the discussion leading to Equation (7.5), we write the secular equation for the electromagnetic field tensor in the form

$$\det (F_n^k - \lambda \delta_n^k) = 0, \tag{15.58}$$

where λ is by definition a scalar invariant. Using here (15.27) and introducing for simplicity a vector $\vec{e} = \vec{E}/c$, we obtain

$$\begin{vmatrix} -\lambda & e_x & e_y & e_z \\ e_x & -\lambda & B_z & -B_y \\ e_y & -B_z & -\lambda & B_x \\ e_z & B_y & -B_x & -\lambda \end{vmatrix} = 0. \tag{15.59}$$

Expanding the expression (15.59) and adding together the terms of the same order in the scalar parameter λ, we obtain

$$\begin{aligned} &- \lambda[-\lambda(\lambda^2 + B_x^2) - B_z(\lambda B_z - B_x B_y) - B_y(B_z B_x + \lambda B_y)] \\ &- e_x[e_x(\lambda^2 + B_x^2) - B_z(-\lambda e_y - B_x e_z) - B_y(-e_y B_x + \lambda e_z)] \\ &+ e_y[e_x(\lambda B_z - B_x B_y) + \lambda(-\lambda e_y - B_x e_z) - B_y(e_y B_y + e_z B_z)] \\ &- e_z[e_x(B_z B_x + \lambda B_y) + \lambda(-e_y B_x + \lambda e_z) + B_z(e_y B_y + e_z B_z)] = 0 \end{aligned} \tag{15.60}$$

or

$$\begin{aligned} &\lambda^4 + \lambda^2 B_x^2 + \lambda^2 B_z^2 - \lambda B_x B_y B_z + \lambda^2 B_y^2 + \lambda B_x B_y B_z \\ &- \lambda^2 e_x^2 - e_x^2 B_x^2 - \lambda e_x e_y B_z - e_x B_x e_z B_z - e_x B_x e_y B_y + \lambda e_x B_y e_z \\ &+ \lambda e_x e_y B_z - e_x B_x e_y B_y - \lambda^2 e_y^2 - \lambda B_x e_y e_z - e_y^2 B_y^2 - e_y B_y e_z B_z \\ &- e_x B_x e_z B_z - \lambda e_x B_y e_z + \lambda B_x e_y e_z - \lambda^2 e_z^2 - e_y B_y e_z B_z - e_z^2 B_z^2 = 0. \end{aligned} \tag{15.61}$$

After regrouping and cancellation of all terms equal in magnitude but with opposite signs, we obtain

$$\begin{aligned} &\lambda^4 + \lambda^2 \left(B_x^2 + B_y^2 + B_z^2 - e_x^2 - e_y^2 - e_z^2 \right) - \left(e_x^2 B_x^2 + e_y^2 B_y^2 \right. \\ &\left. + e_z^2 B_z^2 + 2 e_x B_x e_y B_y + 2 e_x B_x e_z B_z + 2 e_y B_y e_z B_z \right) = 0. \end{aligned} \tag{15.62}$$

In the three-dimensional vector notation, (15.62) becomes

$$\lambda^4 + \lambda^2 (B^2 - e^2) - (\vec{e} \cdot \vec{B})^2 = 0. \tag{15.63}$$

Using here $\vec{e} = \vec{E}/c$, we obtain

$$\lambda^4 + \lambda^2 \left(B^2 - \frac{E^2}{c^2} \right) - \left(\frac{\vec{E} \cdot \vec{B}}{c} \right)^2 = 0. \tag{15.64}$$

Equation (15.64) can also be written in terms of the electromagnetic field tensor F_{kn} as follows:

$$\lambda^4 + \lambda^2 \left(\frac{1}{2} F_{kn} F^{kn} \right) - \left(\frac{1}{8} e^{jkln} F_{jk} F_{ln} \right)^2 = 0. \tag{15.65}$$

Since λ is an invariant absolute scalar, the quantity

$$\frac{1}{2} F_{kn} F^{kn} = B^2 - \frac{E^2}{c^2} = \text{Invariant} \tag{15.66}$$

is an absolute invariant of the electromagnetic field. The quantity

$$\frac{1}{8} e^{jkln} F_{jk} F_{ln} = \frac{\vec{E} \cdot \vec{B}}{c} = \text{Relative invariant} \tag{15.67}$$

is a relative invariant, since $\vec{B} = \text{curl} \vec{A}$ is a relative axial vector. However, from (15.65) we see that the square of the quantity (15.67) is also an absolute invariant of the electromagnetic field. These are the only two scalar invariants that can be constructed using the electromagnetic field tensor F_{kn}.

► Chapter 16

Electromagnetic Field Equations

16.1 Electromagnetic Current Vector

Let us consider a system consisting of a number of charged particles moving in an electromagnetic field specified by the electromagnetic potential four-vector

$$A_n = A_n(x^k) \quad (k, n = 0, 1, 2, 3). \tag{16.1}$$

As the action of the system is an additive quantity, the total action of the interaction between the charged particles and the electromagnetic field is a sum of terms (15.7), given by

$$I_Q = - \sum_M q_M \int_{S_A}^{S_B} A_n dx^n, \tag{16.2}$$

where M is just a label for the Mth particle and not a tensor index. An explicit summation sign is therefore required in Equation (16.2). In electromagnetic field theory it is usually assumed that there is a continuous distribution of charges in three-dimensional space, with the charge density defined by

$$\sigma = \frac{dq}{dV}, \tag{16.3}$$

where the differential dV is an infinitesimal volume element of the three-dimensional space. In such a case Equation (16.2) can be written as follows:

$$I_Q = -\int_V \sigma \, dV \int_{S_A}^{S_B} A_n \frac{dx^n}{dt} \frac{dx^0}{c}. \tag{16.4}$$

Let us now introduce the four-dimensional volume element, defined by the expression

$$d\Omega = dx^0 \, dV = dx^0 \, dx^1 \, dx^2 \, dx^3. \tag{16.5}$$

Since four-dimensional space in the special theory of relativity is a pseudo-Euclidean space, the four-dimensional volume element (16.5) is a scalar invariant. Substituting (16.5) into (16.4), we obtain

$$I_Q = -\frac{1}{c} \int_\Omega \sigma \frac{dx^n}{dt} A_n \, d\Omega. \tag{16.6}$$

Since the action integral (16.6) and the four-dimensional volume element (16.5) are both scalar invariants and A_n is a covariant four-vector, the system defined by

$$J^n = \sigma \frac{dx^n}{dt}, \tag{16.7}$$

is a contravariant four-vector called the *electromagnetic current vector*. The temporal component of the four-vector (16.7) is proportional to the charge density σ and is given by

$$J^0 = c\sigma. \tag{16.8}$$

The spatial components of the four-vector (16.7) constitute a three-dimensional *current density vector*. It is the flux of the charge q through the element of the surface Π that surrounds the volume domain V of the three-dimensional space in which the charges are distributed.

$$J^\alpha = \sigma v^\alpha = \frac{d^3 q}{d\Pi \, dt}. \tag{16.9}$$

The components of the electromagnetic current vector can therefore be structured in the form

$$J^n = \{c\sigma, J_x, J_y, J_z\}$$
$$J_n = g_{nk} J^k = \{c\sigma, -J_x, -J_y, -J_z\} \tag{16.10}$$

Substituting (16.7) into (16.6) we obtain

$$I_Q = -\frac{1}{c} \int_\Omega J^n A_n \, d\Omega = \frac{1}{c} \int_\Omega \mathcal{L}_Q \, d\Omega, \quad \mathcal{L}_Q = -J^n A_n, \tag{16.11}$$

where $\mathcal{L}_Q$ is the *interaction Lagrangian density*. Because of the electric-charge conservation law, the charge density σ and the current density $\vec{J}$ satisfy the *continuity equation*. According to the electric-charge conservation law, the negative increment of the electric charge q within a three-dimensional volume V is equal to the total flux of electric charges through the boundary surface Π of the volume V in the unit of time. Thus in the three-dimensional vector notation, we may write

$$-\frac{d}{dt}\int_V \sigma\,dV = \int_\Pi \vec{J}\cdot d\vec{\Pi}. \tag{16.12}$$

Using the three-dimensional Gauss theorem and regrouping, we obtain

$$\int_V \left(\frac{\partial\sigma}{\partial t} + \operatorname{div}\vec{J}\right)dV = 0. \tag{16.13}$$

From (16.14), we obtain the differential continuity equation in the form

$$\frac{\partial\sigma}{\partial t} + \operatorname{div}\vec{J} = 0. \tag{16.14}$$

In tensor notation, the result (16.14) becomes

$$\frac{\partial J^n}{\partial x^n} = 0 \quad (n = 0, 1, 2, 3). \tag{16.15}$$

The continuity equation (16.15) is therefore related to the conservation law of the total charge in the entire system, defined by

$$q = \int_V \sigma\,dV = \frac{1}{c}\int_V J^0\,dV, \tag{16.16}$$

where V is the entire available volume of the system. Since there is no flux of electric charges through the boundary surface Π of the volume V of the entire system, the surface integral in (16.12) vanishes and we see that the time derivative of the quantity (16.16) is equal to zero.

16.2 Maxwell Equations

The objective of this section is to derive the differential equations satisfied by the electromagnetic field tensor and its components. From the definition of the electromagnetic field tensor (15.14), i.e.,

$$F_{kn} = \frac{\partial A_n}{\partial x^k} - \frac{\partial A_k}{\partial x^n} = \partial_k A_n - \partial_n A_k, \tag{16.17}$$

we see that it satisfies the cyclic equation

$$\partial_j F_{kn} + \partial_n F_{jk} + \partial_k F_{nj} = \partial_j \partial_k A_n - \partial_j \partial_n A_k$$
$$+ \partial_n \partial_j A_k - \partial_n \partial_k A_j + \partial_k \partial_n A_j - \partial_k \partial_j A_n = 0 \qquad (16.18)$$

Thus the first differential equation satisfied by the electromagnetic field tensor is

$$\frac{\partial F_{kn}}{\partial x^j} + \frac{\partial F_{jk}}{\partial x^n} + \frac{\partial F_{nj}}{\partial x^k} = 0. \qquad (16.19)$$

When all three indices j, k, and n are equal to each other ($j = k = n$), Equation (16.19) is a trivial identity since $F_{kn} = 0$ for $k = n$. When two of the indices j, k, and n are equal to each other ($j = k$ or $j = n$ or $k = n$), Equation (16.19) is also a trivial identity due to the antisymmetry of the electromagnetic field tensor $F_{kn} = -F_{nk}$. The only equations of interest are the four equations obtained for $j \neq k \neq n$. Thus we may write

$$\frac{\partial F_{01}}{\partial x^2} + \frac{\partial F_{20}}{\partial x^1} + \frac{\partial F_{12}}{\partial x^0} = 0$$

$$\frac{\partial F_{01}}{\partial x^3} + \frac{\partial F_{30}}{\partial x^1} + \frac{\partial F_{13}}{\partial x^0} = 0$$

$$\frac{\partial F_{02}}{\partial x^3} + \frac{\partial F_{30}}{\partial x^2} + \frac{\partial F_{23}}{\partial x^0} = 0 \qquad (16.20)$$

$$\frac{\partial F_{12}}{\partial x^3} + \frac{\partial F_{31}}{\partial x^2} + \frac{\partial F_{23}}{\partial x^1} = 0.$$

Using here the components of the covariant electromagnetic field tensor (15.25), we obtain

$$\frac{\partial E_x}{\partial y} - \frac{\partial E_y}{\partial x} + \frac{\partial B_z}{\partial t} = 0$$

$$\frac{\partial E_x}{\partial z} - \frac{\partial E_z}{\partial x} + \frac{\partial B_y}{\partial t} = 0$$

$$\frac{\partial E_y}{\partial z} - \frac{\partial E_z}{\partial y} + \frac{\partial B_x}{\partial t} = 0 \qquad (16.21)$$

$$\frac{\partial B_z}{\partial z} + \frac{\partial B_y}{\partial y} + \frac{\partial B_x}{\partial x} = 0.$$

In the three-dimensional vector notation the result (16.21) gives the first pair of *Maxwell equations*, i.e.,

$$\text{curl } \vec{E} + \frac{\partial \vec{B}}{\partial t} = 0, \quad \text{div } \vec{B} = 0. \qquad (16.22)$$

Thus Equation (16.19) is the four-dimensional form of the first pair of Maxwell equations.

In order to derive the second pair of Maxwell equations we need to define the action of the electromagnetic field. The total action for a system consisting of a continous distribution of charged particles in the electromagnetic field is given by

$$I = I_S + I_Q + I_F, \tag{16.23}$$

where I_S is the action for the free particle distribution that does not include the electromagnetic fields or potentials. Its explicit form is therefore not needed in the present section. The term I_Q is the action describing the interaction of the charge distribution with the electromagnetic field and it is given by (16.11). The term I_F is the action of the free electromagnetic field that is an integral of an invariant scalar function called the *Lagrangian density* $\mathcal{L}_F$ over the entire three-dimensional volume V in the time interval $[t_A, t_B]$. Thus the action of the electromagnetic field is of the form

$$I_F = \int_V \int_{t_A}^{t_B} \mathcal{L}_F(A_n, \partial_k A_n) \, dV \, dt = \frac{1}{c} \int_\Omega \mathcal{L}_F(A_n, \partial_k A_n) \, d\Omega. \tag{16.24}$$

The invariant Lagrangian density function $\mathcal{L}_F$ cannot depend on the potentials A_n, as they are not uniquely defined. It may only depend on the space-time derivatives of the potentials $\partial_k A_n$, or in other words on the electromagnetic field tensor F_{kn}. Thus we may write

$$I_F = \frac{1}{c} \int_\Omega \mathcal{L}_F(\partial_k A_n) \, d\Omega = \frac{1}{c} \int_\Omega \mathcal{L}_F(F_{kn}) \, d\Omega. \tag{16.25}$$

Furthermore, because of the linearity of the electromagnetic field equations, the invariant Lagrangian density function $\mathcal{L}_F$ must be at most a quadratic function of the electromagnetic field tensor. The only field invariant that satisfies this condition is given by (15.66). The invariant Lagrangian density $\mathcal{L}_F$ is therefore given by

$$\mathcal{L}_F(F_{kn}) = -\frac{1}{4\mu_0} F^{kn} F_{kn} = \frac{\epsilon_0 E^2}{2} - \frac{B^2}{2\mu_0}. \tag{16.26}$$

The invariant (15.66) can be multiplied by an arbitrary constant, and we have chosen this constant to be equal to $(-2\mu_0)^{-1}$ in order to secure the correct physical dimensions. In (16.26) the quantities μ_0 and $\epsilon_0 = (\mu_0 c^2)^{-1}$ are the vacuum magnetic permeability and the vacuum electric permitivity, respectively. Substituting (16.26) into (16.25) we obtain

$$I_F = -\frac{1}{4c\mu_0} \int_\Omega F^{kn} F_{kn} \, d\Omega = \int_V \int_{t_A}^{t_B} \left(\frac{\epsilon_0 E^2}{2} - \frac{B^2}{2\mu_0} \right) dV \, dt. \quad (16.27)$$

Substituting (16.11) and (16.27) into (16.23) we obtain the total action of the system in the form

$$I = I_S - \frac{1}{c\mu_0} \int_\Omega \left(\mu_0 J^n A_n + \frac{1}{4} F^{kn} F_{kn} \right) d\Omega = \frac{1}{c} \int_\Omega \mathcal{L} \, d\Omega. \quad (16.28)$$

During the derivation of the equations of motion of a charged particle in the electromagnetic field (15.16) in the previous chapter, we assumed that the electromagnetic field is defined by four given functions forming the four-vector $A_n = A_n(x^k)$. We have therefore only varied the Lagrangian (15.9) with respect to the quantities describing the motion of the particle, i.e., the space-time coordinates x^k and coordinates of the four-velocity u^k. On the other hand, in the present derivation of the electromagnetic field equations, we assume that the motion of the charged particles or a continuous distribution of charges in space is defined by given space-time coordinates x^k and coordinates of the four-velocity u^k. We will therefore calculate the variations of the action (16.28) with respect to the quantities describing the electromagnetic field, i.e., the electromagnetic potential four-vector A_n and the electromagnetic field tensor F_{kn}. Thus the variation of the first term in (16.28) is zero and this term does not contribute to the derivation of the electromagnetic field equations. It can therefore be omitted from the action of the system. Furthermore, the electromagnetic current four-vector J^n, as a function of a given charge distribution and its motion in space, is also not varied. The variation of the action (16.28) is then given by

$$\delta I = \frac{1}{c} \int_\Omega \delta \mathcal{L} \, d\Omega = 0, \quad (16.29)$$

with

$$\mathcal{L} (A_n, \partial_k A_n) = -J^n A_n - \frac{1}{4\mu_0} F^{kn} F_{kn}. \quad (16.30)$$

From Equation (16.29) we may write

$$\delta I = \frac{1}{c} \int_\Omega \left[\frac{\partial \mathcal{L}}{\partial A_n} \delta A_n + \frac{\partial \mathcal{L}}{\partial (\partial_k A_n)} \delta (\partial_k A_n) \right] d\Omega = 0. \quad (16.31)$$

Since we have $\delta(\partial_k A_n) = \partial_k(\delta A_n)$, we may write

$$\delta I = \frac{1}{c} \int_\Omega \left\{ \frac{\partial \mathcal{L}}{\partial A_n} - \partial_k \left[\frac{\partial \mathcal{L}}{\partial (\partial_k A_n)} \right] \right\} \delta A_n \, d\Omega$$

$$+ \frac{1}{c} \int_\Omega \partial_k \left[\frac{\partial \mathcal{L}}{\partial (\partial_k A_n)} \delta A_n \right] d\Omega = 0. \tag{16.32}$$

Applying the Gauss theorem to the second integral on the right-hand side of Equation (16.32), it is reduced to the integral over the hypersurface that encloses the given domain Ω of the four-dimensional space-time. On the other hand, the variation of the electromagnetic field variables is assumed to be zero on the boundary of the domain Ω, and this integral vanishes. Equation (16.32) then becomes

$$\delta I = \frac{1}{c} \int_\Omega \left\{ \frac{\partial \mathcal{L}}{\partial A_n} - \partial_k \left[\frac{\partial \mathcal{L}}{\partial(\partial_k A_n)} \right] \right\} \delta A_n \, d\Omega = 0. \tag{16.33}$$

Since the variation δA_n is arbitrary, from (16.33) we obtain the electromagnetic field equations in the following form:

$$\frac{\partial \mathcal{L}}{\partial A_n} - \partial_k \left[\frac{\partial \mathcal{L}}{\partial(\partial_k A_n)} \right] = 0. \tag{16.34}$$

Now, using the expression (16.30), we obtain

$$\frac{\partial \mathcal{L}}{\partial A_n} = -J^n, \tag{16.35}$$

and

$$\frac{\partial \mathcal{L}}{\partial (\partial_k A_n)} = \frac{\partial \mathcal{L}}{\partial F_{jl}} \frac{\partial F_{jl}}{\partial (\partial_k A_n)} = \frac{\partial \mathcal{L}}{\partial F_{jl}} \frac{\partial}{\partial (\partial_k A_n)} (\partial_j A_l - \partial_l A_j)$$

$$= \frac{\partial \mathcal{L}}{\partial F_{jl}} \left(\delta_j^k \delta_l^n - \delta_l^k \delta_j^n \right) = \frac{\partial \mathcal{L}}{\partial F_{kn}} - \frac{\partial \mathcal{L}}{\partial F_{nk}} = 2 \frac{\partial \mathcal{L}}{\partial F_{kn}}$$

$$= 2 \frac{\partial}{\partial F_{kn}} \left(-\frac{1}{4\mu_0} F^{jl} F_{jl} \right) = -\frac{1}{2\mu_0} \frac{\partial}{\partial F_{kn}} \left(F^{jl} F_{jl} \right)$$

$$= -\frac{1}{2\mu_0} F^{jl} \left(\delta_j^k \delta_l^n - \delta_l^k \delta_j^n \right) = -\frac{1}{2\mu_0} (F^{kn} - F^{nk}) = -\frac{1}{\mu_0} F^{kn} \tag{16.36}$$

Substituting (16.35) and (16.36) into the field Equations (16.34), we obtain

$$-J^n - \partial_k \left(-\frac{1}{\mu_0} F^{kn} \right) = 0, \tag{16.37}$$

or, after regrouping,

$$\frac{\partial F^{kn}}{\partial x^k} = \mu_0 J^n \quad (k, n = 0, 1, 2, 3).$$ (16.38)

The temporal component of Equation (16.38) gives

$$\frac{\partial F^{0\alpha}}{\partial x^\alpha} = \frac{1}{c}\frac{\partial E^\alpha}{\partial x^\alpha} = \frac{1}{c}\operatorname{div}\vec{E} = \mu_0 J^0 = \mu_0 c \sigma,$$ (16.39)

or

$$\operatorname{div}\vec{E} = \mu_0 c^2 \sigma = \frac{\sigma}{\epsilon_0}.$$ (16.40)

The three spatial Equations (16.38) give

$$\frac{\partial F^{0\alpha}}{\partial x^0} + \frac{\partial F^{\beta\alpha}}{\partial x^\beta} = \mu_0 J^\alpha,$$ (16.41)

or

$$-\frac{1}{c^2}\frac{\partial E^\alpha}{\partial t} - e^{\beta\alpha\omega}\frac{\partial B_\omega}{\partial x^\beta} = -\frac{1}{c^2}\frac{\partial E^\alpha}{\partial t} + e^{\alpha\beta\omega}\frac{\partial B_\omega}{\partial x^\beta} = \mu_0 J^\alpha.$$ (16.42)

In the three-dimensional vector notation Equation (16.42) is given by

$$\operatorname{curl}\vec{B} - \frac{1}{c^2}\frac{\partial \vec{E}}{\partial t} = \mu_0 \vec{J}.$$ (16.43)

The Equations (16.40) and (16.43) are the second pair of Maxwell equations. Thus the complete system of electromagnetic field equations in the four-dimensional notation is given by

$$\frac{\partial F_{kn}}{\partial x^j} + \frac{\partial F_{jk}}{\partial x^n} + \frac{\partial F_{nj}}{\partial x^k} = 0, \quad \frac{\partial F^{kn}}{\partial x^k} = \mu_0 J^n,$$ (16.44)

with the continuity equation

$$\frac{\partial J^n}{\partial x^n} = 0.$$ (16.45)

The four-dimensional formulation of Maxwell Equations (16.44) with (16.45) is quite compact and it nicely emphasizes the relativistic nature of these equations.

16.3 Electromagnetic Potentials

The objective of the present section is to formulate the differential equations for the electromagnetic potentials $A_n = A_n(x^k)$ and to outline their solutions. In order to derive the differential equations for the electromagnetic potentials, we substitute the definition of the electromagnetic field tensor (15.14) into Equation (16.38). Thus we obtain

$$\partial_k(\partial^k A^n - \partial^n A^k) = \partial_k \partial^k A^n - \partial^n \partial_k A^k = \mu_0 J^n. \tag{16.46}$$

Using the Lorentz gauge (15.45), the second term on the left-hand side vanishes, and we obtain the differential equation for the potentials of the electromagnetic field in the form

$$\frac{\partial^2 A^n}{\partial x^k \partial x_k} = \mu_0 J^n. \tag{16.47}$$

By expanding Equation (16.47) we obtain

$$\frac{1}{c^2}\frac{\partial^2 A^n}{\partial t^2} - \nabla^2 A^n = \mu_0 J^n. \tag{16.48}$$

or

$$\nabla^2 A^n - \frac{1}{c^2}\frac{\partial^2 A^n}{\partial t^2} = -\mu_0 J^n. \tag{16.49}$$

The solution of this equation has the form

$$A^n(x^k) = \frac{\mu_0}{4\pi}\int_{\bar{V}}\frac{J^n(x^0 - R, x^\alpha)}{R}d\bar{V}, \tag{16.50}$$

where the integral is over the domain $\bar{V}$ where the sources of the electromagnetic field are distributed. The quantity R in (16.50) is the three-dimensional distance between the position of the sources and the position where the potential is calculated, i.e.,

$$R = \sqrt{|x^\alpha - \bar{x}^\alpha|^2}. \tag{16.51}$$

Thus, for a given electromagnetic current four-vector $J^n(x^k)$, we can calculate the electromagnetic potential four-vector $A^n(x^k)$ using the result (16.50). The electromagnetic field tensor is then obtained using the definition (15.14).

16.4 Energy–Momentum Tensor

We have shown earlier that the momentum conservation law is a consequence of the homogeneity of space and the energy conservation law is a consequence of the homogeneity of time. In the four-vector language, due to the homogeneity of space-time, the physical properties of a free field remain unchanged with respect to the space-time translations. Thus, the energy–momentum tensor of the electromagnetic field is a field invariant defined with respect to the space-time translations. The Lagrangian of the free electromagnetic field is given by (16.26), i.e.,

$$\mathcal{L}(F_{kn}) = \mathcal{L}_F(F_{kn}) = -\frac{1}{4\mu_0} F^{kn} F_{kn}. \tag{16.52}$$

Furthermore, it is assumed that there are no sources of the electromagnetic field, i.e., that the electromagnetic current four-vector is equal to zero, $J^n = 0$. Thus the interaction Lagrangian $\mathcal{L}_Q$, defined by (16.11), vanishes. The second pair of Maxwell Equations (16.38) then gives

$$\partial_k F^{kn} = 0. \tag{16.53}$$

In order to define the energy–momentum tensor of the electromagnetic field, let us first calculate the space-time derivatives of the Lagrangian density $\mathcal{L}$ as follows:

$$\partial_n \mathcal{L} = \frac{\partial \mathcal{L}}{\partial A_j} \partial_n A_j + \frac{\partial \mathcal{L}}{\partial(\partial_k A_j)} \partial_n(\partial_k A_j). \tag{16.54}$$

Using the electromagnetic field Equations (16.34), we obtain

$$\partial_n \mathcal{L} = \partial_k \left[\frac{\partial \mathcal{L}}{\partial(\partial_k A_j)} \right] \partial_n A_j + \frac{\partial \mathcal{L}}{\partial(\partial_k A_j)} \partial_k(\partial_n A_j), \tag{16.55}$$

or

$$\partial_n \mathcal{L} = \partial_k \left[\frac{\partial \mathcal{L}}{\partial(\partial_k A_j)} \partial_n A_j \right]. \tag{16.56}$$

On the other hand, by definition we may write

$$\partial_n \mathcal{L} = \delta_n^k \partial_k \mathcal{L}. \tag{16.57}$$

From (16.56) and (16.57) we obtain

$$\partial_k \left[\frac{\partial \mathcal{L}}{\partial(\partial_k A_j)} \partial_n A_j - \delta_n^k \mathcal{L} \right] = \partial_k T_n^k = 0, \tag{16.58}$$

where T_n^k is the mixed *energy–momentum tensor* of the electromagnetic field, defined by

$$T_n^k = \frac{\partial \mathcal{L}}{\partial(\partial_k A_j)} \partial_n A_j - \delta_n^k \mathcal{L}. \tag{16.59}$$

Substituting (16.36) and (16.52) into (16.59) we obtain

$$T_n^k = -\frac{1}{\mu_0}F^{kj}\partial_n A_j + \frac{1}{4\mu_0}\delta_n^k F^{jl}F_{jl}. \tag{16.60}$$

The result (16.60) can be put into a more convenient form by writing

$$F^{kj}\partial_n A_j = F^{kj}(\partial_n A_j - \partial_j A_n) + F^{kj}\partial_j A_n, \tag{16.61}$$

or

$$F^{kj}\partial_n A_j = F^{kj}F_{nj} + \partial_j(F^{kj}A_n) - A_n\partial_j F^{kj}. \tag{16.62}$$

The third term on the right-hand side of Equation (16.62) vanishes because of the result (16.53). The second term on the right-hand side of Equation (16.61) makes no contribution to the conservation law (16.58), since we have

$$\partial_k\partial_j(F^{kj}A_n) = \partial_k\partial_j V_n^{kj} \equiv 0. \tag{16.63}$$

The expression (16.63) vanishes as a product of the symmetric tensor $\partial_k\partial_j$ and the tensor V_n^{kj} that is antisymmetric with respect to its two upper indices.

Thus the second term does not contribute to the field invariants obtained from the energy–momentum tensor (16.60). Therefore we may replace $F^{kj}\partial_n A_j$ by $F^{kj}F_{nj}$ in (16.60) to obtain the final result for the energy–momentum tensor of the electromagnetic field in the form

$$T_n^k = -\frac{1}{\mu_0}F^{kj}F_{nj} + \frac{1}{4\mu_0}\delta_n^k F^{jl}F_{jl}. \tag{16.64}$$

The contravariant energy–momentum tensor is then given by

$$T^{nk} = -\frac{1}{\mu_0}F^{kj}F_j^n + \frac{1}{4\mu_0}g^{nk}F^{jl}F_{jl}, \tag{16.65}$$

and the covariant energy–momentum tensor is given by

$$T_{nk} = -\frac{1}{\mu_0}F_k^j F_{nj} + \frac{1}{4\mu_0}g_{nk}F^{jl}F_{jl}. \tag{16.66}$$

Using the differential form of the energy–momentum conservation law (16.58), we may write down the integral form as follows:

$$\int_V \partial_k T^{nk}\,dV = \frac{1}{c}\int_V \frac{\partial T^{n0}}{\partial t}\,dV + \int_V \partial_\alpha T^{\beta\alpha}\,dV = 0. \tag{16.67}$$

Applying the three-dimensional Gauss theorem to the second integral in (16.67), we obtain

$$\int_V \partial_k T^{nk} \, dV = \frac{1}{c} \frac{d}{dt} \left[\int_V T^{n0} \, dV \right] + \oint_\Pi T^{\beta\alpha} \, d\Pi_\alpha = 0 \qquad (16.68)$$

where Π is the closed surface surrounding the volume V. If we let the field domain V grow to infinity $(V \to \infty)$, the surface integral in (16.67) vanishes and we have

$$\int_V \partial_k T^{nk} \, dV = \frac{d}{dt} \left[\frac{1}{c} \int_V T^{n0} \, dV \right] = \frac{dp^n}{dt} = 0. \qquad (16.69)$$

Thus we obtain the definition of the conserved *energy–momentum vector* of the electromagnetic field:

$$p^n = \frac{1}{c} \int_V T^{n0} \, dV = \text{Constant}. \qquad (16.70)$$

The energy of the electromagnetic field is then

$$\mathcal{E} = cp^0 = \int_V T^{00} \, dV, \qquad (16.71)$$

and the components of the three-dimensional momentum of the electromagnetic field are given by

$$p^\alpha = \frac{1}{c} \int_V T^{\alpha 0} \, dV = \text{Constant}. \qquad (16.72)$$

In order to calculate the energy (16.71) we first use (16.65) with (16.27) to calculate

$$T^{00} = -\frac{1}{\mu_0} F^{0\alpha} F^0_\alpha - \left(\frac{\epsilon_0 E^2}{2} - \frac{B^2}{2\mu_0} \right). \qquad (16.73)$$

Using the definitions of the mixed and contravariant electromagnetic field tensors, (15.27) and (15.29), respectively, we obtain

$$T^{00} = -\frac{1}{\mu_0} \left(-\frac{E^2}{c^2} \right) - \left(\frac{\epsilon_0 E^2}{2} - \frac{B^2}{2\mu_0} \right)$$

$$= \epsilon_0 E^2 - \left(\frac{\epsilon_0 E^2}{2} - \frac{B^2}{2\mu_0} \right), \qquad (16.74)$$

or

$$T^{00} = \frac{\epsilon_0 E^2}{2} + \frac{B^2}{2\mu_0}. \qquad (16.75)$$

Substituting (16.75) into (16.71) we obtain the expression for the total energy of the electromagnetic field,

$$\mathcal{E} = \int_V \left(\frac{\epsilon_0 E^2}{2} + \frac{B^2}{2\mu_0} \right) dV = \int_V \varepsilon \, dV, \tag{16.76}$$

where ε is the electromagnetic energy density given by

$$\varepsilon = T^{00} = \frac{\epsilon_0 E^2}{2} + \frac{B^2}{2\mu_0}. \tag{16.77}$$

In order to calculate the momentum (16.72) we first use (16.65) with (16.27) to calculate

$$T^{\alpha 0} = -\frac{1}{\mu_0} F^{0j} F_j^\alpha - g^{\alpha 0} \left(\frac{\epsilon_0 E^2}{2} - \frac{B^2}{2\mu_0} \right). \tag{16.78}$$

Using $g^{\alpha 0} = 0$, we obtain

$$T^{\alpha 0} = -\frac{1}{\mu_0} F^{0\beta} F_\beta^\alpha = -\frac{1}{\mu_0} F^{0\beta} g_{\beta \nu} F^{\nu \alpha} = -\frac{1}{\mu_0} F_\nu^0 F^{\alpha \nu}. \tag{16.79}$$

Using (15.20) in the form

$$F_\nu^0 = -\frac{1}{c} E_\nu, \quad F^{\nu \alpha} = -e^{\nu \alpha \omega} B_\omega, \tag{16.80}$$

we obtain

$$T^{\alpha 0} = -\frac{1}{c\mu_0} e^{\nu \alpha \omega} E_\nu B_\omega = +\frac{1}{c\mu_0} e^{\alpha \nu \omega} E_\nu B_\omega. \tag{16.81}$$

Here, using the notation $\vec{D} = \epsilon_0 \vec{E}$ and $(c\mu_0)^{-1} = c\epsilon_0$, the result (16.81) becomes

$$T^{\alpha 0} = c(\vec{D} \times \vec{B})^\alpha. \tag{16.82}$$

Substituting (16.82) into (16.72) we obtain the expression for the total momentum of the electromagnetic field:

$$\vec{p} = \int_V (\vec{D} \times \vec{B}) dV = \int_V \vec{\mathcal{P}} \, dV, \tag{16.83}$$

where $\vec{\mathcal{P}}$ is the electromagnetic momentum density given by

$$\vec{\mathcal{P}} = \frac{1}{c} T^{\alpha 0} = \vec{D} \times \vec{B}. \tag{16.84}$$

In order to calculate the spatial components of the electromagnetic energy–momentum tensor in terms of the three-dimensional electric field vector $\vec{E}$ and magnetic induction vector $\vec{B}$, we use $F^{kj} = -F^{jk}$ and (16.26) to rewrite the definition (16.64) as follows:

$$T_n^k = \frac{1}{\mu_0} \left[F_{nj}F^{jk} - \frac{1}{2}\delta_n^k(e^2 - B^2) \right]$$

$$= \frac{1}{\mu_0} \left[\Theta_n^k - \frac{1}{2}\delta_n^k(e^2 - B^2) \right], \qquad (16.85)$$

where we introduced the vector $\vec{e} = \vec{E}/c$ and the tensor $\Theta_n^k = F_{nj}F^{jk}$ to simplify the calculations. The contravariant electromagnetic energy–momentum tensor is then given by

$$T^{nk} = g^{nj}T_j^k = \frac{1}{\mu_0} \left[\Theta^{nk} - \frac{1}{2}g^{nk}(e^2 - B^2) \right] \qquad (16.86)$$

We can now use the matrix forms of the covariant and contravariant electromagnetic field tensors, (15.25) and (15.29), respectively, to calculate the components of the tensor Θ_n^k:

$$\left[\Theta_n^k \right] = \begin{bmatrix} 0 & e_x & e_y & e_z \\ -e_x & 0 & -B_z & B_y \\ -e_y & B_z & 0 & -B_x \\ -e_z & -B_y & B_x & 0 \end{bmatrix} \begin{bmatrix} 0 & -e_x & -e_y & -e_z \\ e_x & 0 & -B_z & B_y \\ e_y & B_z & 0 & -B_x \\ e_z & -B_y & B_x & 0 \end{bmatrix}$$

$$= \begin{bmatrix} e^2 & e_yB_z - e_zB_y & e_zB_x - e_xB_z & e_xB_y - e_yB_x \\ -e_yB_z + e_zB_y & e_x^2 - B_y^2 - B_z^2 & e_xe_y + B_xB_y & e_xe_z + B_xB_z \\ -e_zB_x + e_xB_z & e_ye_x + B_yB_x & e_y^2 - B_x^2 - B_z^2 & e_ye_z + B_yB_z \\ -e_xB_y + e_yB_x & e_ze_x + B_zB_x & e_ze_y + B_zB_y & e_z^2 - B_x^2 - B_y^2 \end{bmatrix}$$

$$= \begin{bmatrix} e^2 & (\vec{e} \times \vec{B})_x & (\vec{e} \times \vec{B})_y & (\vec{e} \times \vec{B})_z \\ -(\vec{e} \times \vec{B})_x & e_x^2 + B_x^2 - B^2 & e_xe_y + B_xB_y & e_xe_z + B_xB_z \\ -(\vec{e} \times \vec{B})_y & e_ye_x + B_yB_x & e_y^2 + B_y^2 - B^2 & e_ye_z + B_yB_z \\ -(\vec{e} \times \vec{B})_z & e_ze_x + B_zB_x & e_ze_y + B_zB_y & e_z^2 + B_z^2 - B^2 \end{bmatrix}.$$

$$(16.87)$$

The spatial part of the tensor Θ_n^k is then given by the following matrix:

$$\left[\Theta_\beta^\alpha \right] = \begin{bmatrix} e_x^2 + B_x^2 - B^2 & e_xe_y + B_xB_y & e_xe_z + B_xB_z \\ e_ye_x + B_yB_x & e_y^2 + B_y^2 - B^2 & e_ye_z + B_yB_z \\ e_ze_x + B_zB_x & e_ze_y + B_zB_y & e_z^2 + B_z^2 - B^2 \end{bmatrix}. \qquad (16.88)$$

The contravariant components of the three-dimensional system (16.88) are obtained from

$$\Theta^{\alpha\beta} = g^{\beta\nu}\Theta_\nu^\alpha, \qquad (16.89)$$

or, in the matrix form,

$$
[\Theta^{\alpha\beta}] = \begin{bmatrix} -e_x^2 - B_x^2 + B^2 & -e_x e_y - B_x B_y & -e_x e_z - B_x B_z \\ -e_y e_x - B_y B_x & -e_y^2 - B_y^2 + B^2 & -e_y e_z - B_y B_z \\ -e_z e_x - B_z B_x & -e_z e_y - B_z B_y & -e_z^2 - B_z^2 + B^2 \end{bmatrix}.
$$
(16.90)

From the matrix (16.90) we obtain

$$
\Theta^{\alpha\beta} = -e^\alpha e^\beta - B^\alpha B^\beta - g^{\alpha\beta} B^2
$$

$$
= -\frac{E^\alpha E^\beta}{c^2} - B^\alpha B^\beta - g^{\alpha\beta} B^2.
$$
(16.91)

Substituting the result (16.91) into the spatial part of Equation (16.86), we obtain

$$
T^{\alpha\beta} = \frac{1}{\mu_0}\left[-e^\alpha e^\beta - B^\alpha B^\beta - g^{\alpha\beta} B^2 - \frac{1}{2} g^{\alpha\beta}(e^2 - B^2) \right]
$$

$$
= -\frac{1}{\mu_0}\left[e^\alpha e^\beta + B^\alpha B^\beta + \frac{1}{2} g^{\alpha\beta}(e^2 + B^2) \right].
$$
(16.92)

Thus we finally obtain the spatial spatial components of the electromagnetic energy–momentum tensor in terms of the three-dimensional electric field vector $\vec{E}$ and magnetic induction vector $\vec{B}$, as follows:

$$
T^{\alpha\beta} = -\mathcal{T}^{\alpha\beta},
$$
(16.93)

where we define the *Maxwell stress tensor* $\mathcal{T}^{\alpha\beta}$ by the equation

$$
\mathcal{T}^{\alpha\beta} = \epsilon_0 E^\alpha E^\beta + \frac{1}{\mu_0} B^\alpha B^\beta + g^{\alpha\beta}\left(\epsilon_0 E^2 + \frac{B^2}{\mu_0} \right).
$$
(16.94)

Using the results (16.77) with (16.84) and (16.93) we may write the explicit matrix form of the electromagnetic energy–momentum tensor as follows:

$$
[T^{nk}] = \begin{bmatrix} \varepsilon & \mathcal{P}^1 & \mathcal{P}^2 & \mathcal{P}^3 \\ \mathcal{P}^1 & -\mathcal{T}^{11} & -\mathcal{T}^{12} & -\mathcal{T}^{13} \\ \mathcal{P}^2 & -\mathcal{T}^{21} & -\mathcal{T}^{22} & -\mathcal{T}^{23} \\ \mathcal{P}^3 & -\mathcal{T}^{31} & -\mathcal{T}^{32} & -\mathcal{T}^{33} \end{bmatrix}.
$$
(16.95)

From the definition (16.94) we see that the three-dimensional Maxwell stress tensor is a symmetric tensor. By examination of (16.95) we conclude that the complete four-dimensional electromagnetic energy–momentum tensor is also a symmetric tensor, i.e., we have

$$
T^{nk} = T^{kn}, \quad T_{nk} = T_{kn}.
$$
(16.96)

If the electromagnetic field is defined in a domain V with the boundary surface Π, the expression

$$F^\alpha = \int_V \frac{\partial T^{\alpha\beta}}{\partial x^\beta} dV = \oint_\Pi T^{\alpha\beta}\, d\Pi_\beta \qquad (16.97)$$

defines the force due to the electromagnetic field pressure on the boundary surface Π.

Part IV

General Theory of Relativity

▶ Chapter 17

Gravitational Fields

17.1 Introduction

The special theory of relativity discussed in the last four chapters is based on the concept of inertial frames of reference. An inertial frame of reference is defined as a system of reference where a free particle, i.e., a particle on which there is no action of any external forces, moves along a straight line with constant velocity. On the other hand, the gravitational interaction, as one of the fundamental interactions in nature, is a long-range interaction that cannot be screened. The concept of inertial frames of reference is, therefore, not compatible with gravitational phenomena. The only way to define an approximately inertial frame of reference is to visualize it as being far away from any matter. Given the influence of the force of gravity on the observations of various physical phenomena, both on Earth and in the universe as a whole, the concept of inertial frames of reference as a foundation for the formulation of the laws of nature is clearly not sufficient. Since the force of gravity is an unscreened long-range force, it can be considered as an intrinsic property of space-time and related to space-time geometry.

In the absence of matter we can define the inertial frames of reference where the geometry of space-time is pseudo-Euclidean. The space-time metric is then given by

$$ds^2 = g_{kn}dx^k dx^n = c^2 dt^2 - dx^2 - dy^2 - dz^2, \qquad (17.1)$$

such that the components of the metric tensor g_{mn} are constant and given by the following matrix:

$$[g_{mn}] = \begin{bmatrix} 1 & 0 & 0 & 0 \\ 0 & -1 & 0 & 0 \\ 0 & 0 & -1 & 0 \\ 0 & 0 & 0 & -1 \end{bmatrix}. \tag{17.2}$$

In the presence of matter it is not possible to find the four space-time coordinates x^j such that the metric of the space-time manifold is reduced to the pseudo-Euclidean expression (17.1). Thus the geometry of space-time in the presence of matter must be a pseudo-Riemannian geometry, where the components of the metric tensor g_{kn} are functions of the coordinates x^j, i.e., we have

$$g_{kn} = g_{kn}(x^j) \quad (j,k,n = 0,1,2,3). \tag{17.3}$$

The components of the metric tensor of the given space-time manifold (17.3) can therefore be considered as the *gravitation field potentials*; and the gravitational effects are described by the metric itself.

From (17.2) we note that the determinant $g = -1$ of the pseudo-Euclidean metric tensor of special theory of relativity is a negative constant number. In the general theory of relativity $g = g(x^j) < 0$ is a negative definite function of space-time coordinates. In the ordinary positive-definite Riemannian metric spaces, we have frequently used the function $\sqrt{g}$ to define the absolute tensors and invariants. In the pseudo-Euclidean and pseudo-Riemannian space-times of the theory of relativity with $g < 0$, this function is not a real function and it is generally replaced by $\sqrt{-g}$, which is a real function. In the pseudo-Euclidean special theory of relativity we have $\sqrt{-g} = 1$, so the relative tensors and invariants were transformed as the absolute scalars and invariants, and the explicit appearance of the factor $\sqrt{-g}$ was not necessary.

In the presence of gravity it is not possible to reduce the metric of space-time to the pseudo-Euclidean expression (17.1) in the entire space. It is important to note that the gravitational effects are understood as a deviation of the space-time metric from the pseudo-Euclidean metric (17.1). The metric (17.3) is therefore a function of the local distribution of matter as a source of the gravitational field and cannot be fixed arbitrarily in the entire space. The metric tensor, as the gravitation field potential, is in this context a solution of the gravitational field equations, which will be derived in the next chapter.

Although we cannot transform away the gravity in the entire space, we can select the local frames of reference falling freely in the gravitational field where the gravitational effects are locally transformed away. In a small laboratory falling freely in a gravitational field, the laws of nature are therefore the same as those observed in an inertial frame of reference in the absence of gravity. The laboratory has to be small since the gravitational fields are functions of coordinates and vary in magnitude and direction in different points in space. The force of gravity can therefore be considered as uniform and transformed away by a suitable choice of local frame of reference only in a sufficiently small space domain.

Thus we are in most cases able to cover the space-time with a patchwork of *local inertial frames of reference*. The concept of local inertial frames of reference is very useful in the general theory of relativity.

17.2 Time Intervals and Distances

In the general theory of relativity, the choice of the generalized coordinates used to describe the four-dimensional space-time manifold is not restricted in any way. Thus the coordinates x^j are in general not equal to the distances and time intervals between events in the same way as in the special theory of relativity. Therefore, given a set of generalized coordinates x^j, we need to relate these coordinates to the actual distances and time intervals between the observed events.

In the general theory of relativity we denote the proper time by τ. Let us now consider two infinitesimally separated events that take place at exactly the same point in space. Then we have $dx^1 = dx^2 = dx^3 = 0$, and we may write

$$ds^2 = g_{kn}dx^k dx^n = c^2 d\tau^2 = g_{00}\left(dx^0\right)^2. \qquad (17.4)$$

Thus we obtain the relation between the element of the proper time $d\tau$ and the coordinate differential dx^0 as follows:

$$d\tau = \frac{1}{c}\sqrt{g_{00}}\, dx^0, \qquad (17.5)$$

or for the proper time between any two events occuring at the same point in space,

$$\tau = \frac{1}{c}\int \sqrt{g_{00}}\, dx^0. \qquad (17.6)$$

In the special theory of relativity the element dl of spatial distance is defined as the distance between two infinitesimally separated events taking place at the same time. In the general theory of relativity we cannot use this definition, since the proper time is a different function of coordinates at different points in space. In order to find the element dl of the spatial distance, we consider two infinitesimally close points A and B with coordinates x^j and $x^j + dx^j$, respectively. Let us suppose that a light signal is directed from point A to point B and then back from point B to point A along the same path. The time required by the light signal to travel from point A to point B and back, observed from one point in space and multiplied by c, is double the distance between the two points. For the two events representing the departure of the light signal at one point and the arrival at the other point, we know from (13.4) that the square of the interval ds^2 is equal to zero. Thus we may write

$$ds^2 = g_{kn}dx^k dx^n = g_{00}(dx^0)^2 + 2g_{0\alpha}dx^\alpha dx^0 + g_{\alpha\beta}dx^\alpha dx^\beta = 0. \quad (17.7)$$

The equation (17.7) is a quadratic equation with respect to dx^0 and its solutions are given by

$$(dx^0)_\pm = -\frac{g_{0\alpha}dx^\alpha dx^0}{g_{00}} \pm \frac{1}{g_{00}}\sqrt{\left(g_{0\alpha}g_{0\beta} - g_{00}g_{\alpha\beta}\right)dx^\alpha dx^\beta}. \quad (17.8)$$

The coordinate time required by the light signal to travel from point A to point B and back is then given by

$$(dx^0)_+ - (dx^0)_- = \frac{2}{g_{00}}\sqrt{\left(g_{0\alpha}g_{0\beta} - g_{00}g_{\alpha\beta}\right)dx^\alpha dx^\beta}. \quad (17.9)$$

Using (17.5) we obtain the proper time required by the light signal to travel from point A to point B and back, multiplied by c, as follows:

$$cd\tau = \sqrt{g_{00}}\left[(dx^0)_+ - (dx^0)_-\right]$$

$$= \frac{2}{\sqrt{g_{00}}}\sqrt{\left(g_{0\alpha}g_{0\beta} - g_{00}g_{\alpha\beta}\right)dx^\alpha dx^\beta}. \quad (17.10)$$

Thus the spatial element dl between the two points A and B is a half of the proper time interval (17.10) multiplied by c, i.e.,

$$dl = \frac{1}{2}cd\tau = \sqrt{\left(-g_{\alpha\beta} + \frac{g_{0\alpha}g_{0\beta}}{g_{00}}\right)dx^\alpha dx^\beta}. \quad (17.11)$$

Thus the square of the space metric dl^2 can be written in the form

$$dl^2 = \gamma_{\alpha\beta}dx^\alpha dx^\beta \quad \gamma_{\alpha\beta} = -g_{\alpha\beta} + \frac{g_{0\alpha}g_{0\beta}}{g_{00}}. \quad (17.12)$$

It should be noted that the metric tensor g_{kn} in general depends on the time coordinate. The space metric dl^2 is therefore also time dependent, and in general it does not make sense to integrate the element of the spatial distance dl, as such an integral would depend on the world line chosen between the two end points. Only when the metric tensor g_{kn} is not time dependent and the distance can be defined over a finite portion of space does the integral of the element of the spatial distance dl along a space curve have a definite meaning.

17.3 **Particle Dynamics**

In the special theory of relativity the motion of a free particle of mass m is defined by the action integral (14.5). The action integral of a free particle is simply proportional to the arc length in the four-dimensional space-time manifold, i.e.,

$$I = -mc \int_{S_A}^{S_B} ds = -mc \int_{S_A}^{S_B} \sqrt{g_{kn} \frac{dx^k}{ds} \frac{dx^n}{ds}} ds, \qquad (17.13)$$

where m is a mass parameter of the free particle. From (17.13) we see that the Lagrangian function is given by the expression

$$L = -mc \sqrt{g_{kn} u^k u^n}. \qquad (17.14)$$

The equations of motion of a particle in the gravitational field are obtained by variation of the same action integral (17.13), since the introduction of the field of gravity is nothing but the change of the space-time metric to a pseudo-Riemannian metric with a coordinate-dependent metric tensor (17.3). On the other hand, except for the overall constant multiplier $-mc$, the Lagrangian (17.14) is the same as the Lagrangian (11.14) used to derive the geodesic equations (11.25). Thus we conclude that a free particle in the gravitational field moves along the geodesic lines and that the equations of motion are the geodesic equations (11.25), i.e.,

$$\frac{d^2 x^n}{ds^2} + \Gamma^n_{lk} \frac{dx^l}{ds} \frac{dx^k}{ds} = 0 \quad (k, l, n = 0, 1, 2, 3) \qquad (17.15)$$

or

$$\frac{du^n}{ds} + \Gamma^n_{lk} u^l u^k = 0 \quad (k, l, n = 0, 1, 2, 3). \qquad (17.16)$$

In order to find the nonrelativistic limit of Equations (17.16), we first note that the nonrelativistic limit ($v \ll c$) of the components of the four-velocity

(13.39) and (13.40) is given by

$$u^0 = \frac{1}{\sqrt{1 - \frac{v^2}{c^2}}} \approx 1, \quad u^\alpha = \frac{v^\alpha}{c\sqrt{1 - \frac{v^2}{c^2}}} \approx \frac{v^\alpha}{c}. \tag{17.17}$$

Thus the leading term on the right-hand side of the expression (17.16) is the one with $l = k = 0$, such that we may write

$$\frac{du^n}{ds} = -\Gamma^n_{00} \quad (n = 0, 1, 2, 3). \tag{17.18}$$

For a static gravitational field with $\partial_0 g_{00} = 0$, the only nontrivial equations in (17.18) are the spatial equations

$$\frac{du^\alpha}{ds} = -\Gamma^\alpha_{00} \quad (\alpha = 1, 2, 3). \tag{17.19}$$

Using (17.17) and the zeroth-order nonrelativistic approximation for the differential of the parameter $ds \approx c\,dt$, we further obtain

$$\frac{1}{c^2}\frac{dv^\alpha}{dt} = -\Gamma^\alpha_{00} \quad (\alpha = 1, 2, 3). \tag{17.20}$$

Let us now recall the definition of the Christoffel symbols of the first kind (9.28), i.e.,

$$\Gamma_{k,jp} = \frac{1}{2}\left(\frac{\partial g_{kj}}{\partial x^p} + \frac{\partial g_{pk}}{\partial x^j} - \frac{\partial g_{jp}}{\partial x^k}\right). \tag{17.21}$$

For a static gravitational field with $\partial_0 g_{jp} = 0$, we obtain

$$\Gamma_{k,00} = \frac{1}{2}\left(\frac{\partial g_{k0}}{\partial x^0} + \frac{\partial g_{0k}}{\partial x^0} - \frac{\partial g_{00}}{\partial x^k}\right) = -\frac{1}{2}\frac{\partial g_{00}}{\partial x^k}. \tag{17.22}$$

Now using the definition of the Christoffel symbols of the second kind (9.29), i.e.,

$$\Gamma^n_{jp} = g^{nk}\Gamma_{k,jp} = \frac{1}{2}g^{nk}\left(\frac{\partial g_{kj}}{\partial x^p} + \frac{\partial g_{pk}}{\partial x^j} - \frac{\partial g_{jp}}{\partial x^k}\right), \tag{17.23}$$

we obtain

$$\Gamma^n_{00} = g^{nk}\Gamma_{k,00} = -\frac{1}{2}g^{nk}\frac{\partial g_{00}}{\partial x^k}. \tag{17.24}$$

The spatial components of Equation (17.24) give

$$\Gamma^\alpha_{00} = -\frac{1}{2}g^{\alpha k}\frac{\partial g_{00}}{\partial x^k}. \tag{17.25}$$

For a static gravitational field with $\partial_0 g_{00} = 0$ we then obtain

$$\Gamma^\alpha_{00} = -\frac{1}{2} g^{\alpha\beta} \frac{\partial g_{00}}{\partial x^\beta}. \tag{17.26}$$

Substituting (17.26) into (17.20), we obtain

$$\frac{1}{c^2} \frac{dv^\alpha}{dt} = \frac{1}{2} g^{\alpha\beta} \frac{\partial g_{00}}{\partial x^\beta} \quad (\alpha, \beta = 1, 2, 3). \tag{17.27}$$

In order to find the nonrelativistic approximation for the component g_{00} of the metric tensor, let us consider the nonrelativistic Lagrangian for a particle moving in the gravitational field defined by the gravitational potential ϕ, given by

$$L = \mathcal{E}_K - V = \frac{1}{2} mv^2 - m\phi, \quad \phi = \phi(x^\alpha). \tag{17.28}$$

We may always add a constant term to the nonrelativistic Lagrangian without affecting the dynamics. Thus, following the result (14.17) we add a constant $-mc^2$ to the Lagrangian (17.28) and obtain

$$L = -mc^2 + \frac{1}{2} mv^2 - m\phi = -mc^2 - \frac{1}{2} m g_{\alpha\beta} v^\alpha v^\beta - m\phi, \tag{17.29}$$

where $g_{\alpha\beta}$ are the spatial components of the pseudo-Euclidean metric tensor. In the Descartes coordinates they are given by $g_{\alpha\beta} = -\delta_{\alpha\beta}$. Using (17.29) we obtain the nonrelativistic approximation to the action integral (17.13) in the form

$$I \approx -mc \int_{t_A}^{t_B} \left(c + \frac{\phi}{c} + \frac{1}{2} g_{\alpha\beta} \frac{v^\alpha}{c} \frac{dx^\beta}{dt} \right) dt, \tag{17.30}$$

or, after some regrouping,

$$I = -mc \int_{S_A}^{S_B} ds$$

$$\approx -mc \int_{S_A}^{S_B} \left[\left(1 + \frac{\phi}{c^2} \right) c\, dt + \frac{1}{2} g_{\alpha\beta} \frac{v^\alpha}{c} dx^\beta \right]. \tag{17.31}$$

From Equation (17.31) we obtain the nonrelativistic approximation for the line element ds as follows:

$$ds \approx \left(1 + \frac{\phi}{c^2} \right) dx^0 + \frac{1}{2} g_{\alpha\beta} \frac{v^\alpha}{c} dx^\beta. \tag{17.32}$$

The nonrelativistic approximation to the metric of the space-time manifold is then obtained by squaring the expression (17.32) and dropping the terms of the order v^2/c^2 and higher. Thus we obtain

$$ds^2 \approx \left(1 + \frac{\phi}{c^2}\right)^2 (dx^0)^2 + \left(1 + \frac{\phi}{c^2}\right) g_{\alpha\beta} \frac{dx^\alpha}{dt} dx^\beta \frac{dx^0}{c}$$

$$+ \frac{1}{4} \left(g_{\alpha\beta} \frac{v^\alpha}{c} dx^\beta\right)^2. \tag{17.33}$$

Here we may drop the last term as the term of higher order in v^2/c^2 to obtain

$$ds^2 \approx \left(1 + \frac{\phi}{c^2}\right)^2 (dx^0)^2 + \left(1 + \frac{\phi}{c^2}\right) g_{\alpha\beta} dx^\alpha dx^\beta. \tag{17.34}$$

Assuming that the gravitational field is relatively weak, we may use the approximations

$$\left(1 + \frac{\phi}{c^2}\right)^2 \approx 1 + \frac{2\phi}{c^2} \tag{17.35}$$

and

$$\left(1 + \frac{\phi}{c^2}\right) g_{\alpha\beta} dx^\alpha dx^\beta \approx g_{\alpha\beta} dx^\alpha dx^\beta. \tag{17.36}$$

Substituting (17.35) and (17.36) into (17.34), we finally obtain

$$ds^2 \approx \left(1 + \frac{2\phi}{c^2}\right) (dx^0)^2 + g_{\alpha\beta} dx^\alpha dx^\beta. \tag{17.37}$$

From the result (17.37) we obtain the nonrelativistic approximation to the quantity g_{00} in the form

$$g_{00} \approx 1 + \frac{2\phi}{c^2}. \tag{17.38}$$

Substituting (17.38) into (17.27), we obtain

$$\frac{dv^\alpha}{dt} = g^{\alpha\beta} \frac{\partial \phi}{\partial x^\beta} \quad (\alpha, \beta = 1, 2, 3). \tag{17.39}$$

Using here the pseudo-Euclidean spatial metric $g^{\alpha\beta}$, consistent with the result (17.37), we obtain in the three-dimensional vector form

$$\frac{d\vec{v}}{dt} = -\text{grad } \phi, \quad \phi = \phi(\vec{x}). \tag{17.40}$$

If we now recall the nonrelativistic result for the gravitational potential ϕ, we may write

$$\phi(\vec{x}) = -\frac{GM}{r} \Rightarrow \operatorname{grad} \phi = \frac{GM}{r^3}\vec{r}, \qquad (17.41)$$

where M is the mass of the source of the gravitational field, r is the three-dimensional radial coordinate given by

$$r = \sqrt{x^2 + y^2 + z^2}, \qquad (17.42)$$

and G is the gravitational constant given by

$$G = 6.67 \times 10^{-11} \frac{\mathrm{m}^3}{\mathrm{kg\ s}^2}. \qquad (17.43)$$

Substituting (17.41) into (17.40) and multiplying by the mass of the test particle m, we obtain the result for the nonrelativistic gravitational force between two bodies with masses M and m as follows:

$$\vec{F}_G = m\frac{d\vec{v}}{dt} = -\frac{GMm}{r^3}\vec{r}. \qquad (17.44)$$

Thus in the nonrelativistic limit the equations of motion of a particle (17.16) are reduced to the corresponding Newtonian nonrelativistic equations of motion.

17.4 Electromagnetic Field Equations

The objective of the present section is to generalize the electromagnetic field equations, derived in the previous chapter, to the pseudo-Riemannian metric space in the presence of gravity. In the special theory of relativity the electromagnetic field tensor is defined by Equation (15.14), i.e.,

$$F_{kn} = \partial_k A_n - \partial_n A_k \quad (k, n = 0, 1, 2, 3). \qquad (17.45)$$

As the partial derivatives do not transform as vectors in the Riemannian metric spaces, the expression (17.45) has to be modified to the covariant expression

$$F_{kn} = D_k A_n - D_n A_k \quad (k, n = 0, 1, 2, 3). \qquad (17.46)$$

Using the definition of the covariant derivatives (8.24), i.e.,

$$D_k A_n = \partial_k A_n - \Gamma_{nk}^j A_j, \quad D_n A_k = \partial_n A_k - \Gamma_{kn}^j A_j, \qquad (17.47)$$

we obtain

$$F_{kn} = \partial_k A_n - \Gamma^j_{nk} A_j - \partial_n A_k + \Gamma^j_{kn} A_j. \tag{17.48}$$

Using the symmetry of the Christoffel symbols of the second kind (9.9), i.e.,

$$\Gamma^j_{nk} = \Gamma^j_{kn} \quad (k, m, p = 0, 1, 2, 3), \tag{17.49}$$

the second and fourth terms on the right-hand side of Equation (17.47) cancel each other and we reproduce the pseudo-Euclidean result (17.45). Even in the presence of gravity the definition of the electromagnetic field tensor (17.45) remains valid. Using the definition of the covariant derivative of a second-order covariant tensor (8.33), we have

$$D_j F_{kn} = \partial_j F_{kn} - \Gamma^l_{kj} F_{ln} - \Gamma^l_{nj} F_{kl}$$

$$D_n F_{jk} = \partial_n F_{jk} - \Gamma^l_{jn} F_{lk} - \Gamma^l_{kn} F_{jl} \tag{17.50}$$

$$D_k F_{nj} = \partial_k F_{nj} - \Gamma^l_{nk} F_{lj} - \Gamma^l_{jk} F_{nl}.$$

Now, using the antisymmetry of the electromagnetic field tensor

$$F_{ln} = -F_{nl}, \quad F_{kl} = -F_{lk}, \quad F_{jl} = -F_{lj} \tag{17.51}$$

and the symmetry of the Christoffel symbols of the second kind

$$\Gamma^l_{kj} = \Gamma^l_{jk}, \quad \Gamma^l_{nj} = \Gamma^l_{jn}, \quad \Gamma^l_{kn} = \Gamma^l_{nk}, \tag{17.52}$$

we obtain from (17.50) the cyclic expression

$$D_j F_{kn} + D_n F_{jk} + D_k F_{nj} = \partial_j F_{kn} + \partial_n F_{jk} + \partial_k F_{nj} = 0. \tag{17.53}$$

Thus we conclude that the first pair of Maxwell equations remains unchanged in the presence of gravity and has the form

$$\frac{\partial F_{kn}}{\partial x^j} + \frac{\partial F_{jk}}{\partial x^n} + \frac{\partial F_{nj}}{\partial x^k} = 0 \quad (j, k, n = 0, 1, 2, 3). \tag{17.54}$$

In order to derive the second pair of Maxwell equations, let us consider the interaction Lagrangian (16.6) in the form

$$I_Q = -\frac{1}{c} \int_\Omega \sigma \frac{dx^n}{dt} A_n d\Omega. \tag{17.55}$$

The four-dimensional volume element, defined in the pseudo-Euclidean space-time of special theory of relativity by the equation

$$d\Omega = dx^0 dV = dx^0 dx^1 dx^2 dx^3, \tag{17.56}$$

is no longer an absolute scalar. Instead of (17.56) we define the invariant volume element as $\sqrt{-g}\, d\Omega$. Thus we rewrite the interaction Lagrangian

(17.55) as follows:

$$I_Q = -\frac{1}{c}\int_\Omega \frac{c\sigma}{\sqrt{-g}}\frac{dx^n}{dx^0}A_n\sqrt{-g}\,d\Omega = -\frac{1}{c}\int_\Omega J^n A_n\sqrt{-g}\,d\Omega. \quad (17.57)$$

Since the action integral (17.57) and the four-dimensional volume element $\sqrt{-g}\,d\Omega$ are both absolute invariants and A_n is a covariant four-vector, the system defined by

$$J^n = \frac{\sigma c}{\sqrt{-g}}\frac{dx^n}{dx^0} \quad (17.58)$$

is by definition the contravariant electromagnetic current four-vector. The temporal component of the four-vector (17.58) is proportional to the charge density σ and is given by

$$J^0 = c\frac{\sigma}{\sqrt{-g}}, \quad (17.59)$$

and the three-dimensional current density vector is

$$J^\alpha = \frac{\sigma}{\sqrt{-g}}v^\alpha. \quad (17.60)$$

Given the definition (17.58) we can generalize the second pair of Maxwell equations (16.38) by replacing the partial derivative ∂_k by the covariant derivative D_k. The second pair of Maxwell equations then becomes

$$D_k F^{kn} = \mu_0 J^n \quad (k,n = 0,1,2,3). \quad (17.61)$$

Using the definition of divergence of an arbitrary tensor with respect to one of its contravariant indices (10.15), we obtain

$$D_k F^{kn} = \frac{1}{\sqrt{-g}}\frac{\partial}{\partial x^k}\left(\sqrt{-g}F^{kn}\right) = \mu_0 J^n. \quad (17.62)$$

In the same way we generalize the equation of continuity (16.45) to obtain

$$D_n J^n = \frac{1}{\sqrt{-g}}\frac{\partial}{\partial x^n}\left(\sqrt{-g}J^n\right) = 0. \quad (17.63)$$

Thus the complete system of electromagnetic field equations in the pseudo-Riemannian space-time manifold, which includes the gravitational effects, is given by

$$\frac{\partial F_{kn}}{\partial x^j} + \frac{\partial F_{jk}}{\partial x^n} + \frac{\partial F_{nj}}{\partial x^k} = 0, \quad \frac{1}{\sqrt{-g}}\frac{\partial}{\partial x^k}\left(\sqrt{-g}F^{kn}\right) = \mu_0 J^n \quad (17.64)$$

with the continuity equation

$$\frac{1}{\sqrt{-g}}\frac{\partial}{\partial x^n}\left(\sqrt{-g}J^n\right)=0. \tag{17.65}$$

The pseudo-Riemannian four-dimensional formulation of Maxwell equations in the presence of gravity given by (17.64) with (17.65) is quite compact, and it nicely illustrates the general prescription for the introduction of gravity in the laws of physics.

In general, the laws of physics must be valid in all systems of reference and expressed as covariant tensor equations. The field equations of physics, which involve the derivatives of the field variables, must therefore be rewritten such that the ordinary derivatives are replaced by the covariant derivatives. Even if we are working in a flat space-time described by some curvilinear coordinates, we need to use the covariant derivatives if we want the field equations to be valid in all coordinate systems.

If we assume that a physical object described by a set of pseudo-Euclidean equations does not represent an appreciable source of the gravitational field, then it does not significantly influence the components of the metric tensor $g_{kn}(x^j)$ as the potentials of the gravitational field. In such a case the geometry of the four-dimensional space-time manifold is rigidly determined by some massive external sources of the gravitational field. Then the effect of the physical object under consideration on the geometrical structure may be neglected. Under these circumstances the set of pseudo-Euclidean equations describing the physical object can readily be generalized to incorporate the effects of gravity by making the substitution

$$d \rightarrow D, \quad \partial_k \rightarrow D_k, \quad d\Omega \rightarrow \sqrt{-g}\,d\Omega. \tag{17.66}$$

Thus the prescriptions (17.66) can be used to generalize any pseudo-Euclidean equation, used to describe a physical object that is small in comparison to the sources of the gravitational field, to incorporate the gravitational effects and to be valid in all systems of reference.

Chapter 18

Gravitational Field Equations

18.1 The Action Integral

In the previous chapter we concluded that the components of the metric tensor of a given space-time manifold are the potentials of the field of gravity, and the gravitational effects are described by the metric itself. The metric tensor is therefore a system of functions of the local distribution of matter as a source of the gravitational field and cannot be fixed arbitrarily in the entire space. The metric tensor as the field potential is therefore a solution of the gravitational field equations. The objective of the present chapter is to derive the gravitational field equations from the suitable action integral of the gravitational field. The total action integral for a system consisting of a continuous distribution of matter as the source of the gravitational field and the gravitational field itself is given by

$$I = I_G + I_M, \tag{18.1}$$

where I_G is the action of the gravitational field in empty space, where there are no field sources, and I_M is the action describing the interaction of the matter distribution with the gravitational field. The action of the gravitational field in empty space can be written in the form

$$I_G = \frac{1}{c} \int_{\Omega} \mathcal{L}_G(g_{kn}, \partial_j g_{kn}) \sqrt{-g} \, d\Omega, \tag{18.2}$$

where $\mathcal{L}_G$ is the invariant Lagrangian density of the gravitational field that is integrated over the invariant four-dimensional domain Ω. The action describing the interaction of the matter distribution with the gravitational field can be written as follows:

$$I_M = \frac{1}{c} \int_\Omega \mathcal{L}_M \left(g_{kn}, \partial_j g_{kn}\right) \sqrt{-g} d\Omega, \tag{18.3}$$

where $\mathcal{L}_M$ is the invariant Lagrangian density of the source fields. For example, if the source of the gravitational field is the electromagnetic field, then $\mathcal{L}_M$ is the Lagrangian density of the electromagnetic field (16.52), i.e.,

$$\mathcal{L}_M \left(g_{kn}\right) = -\frac{1}{4\mu_0} g_{kj} g_{nl} F^{kn} F^{jl}. \tag{18.4}$$

It should be noted that the electromagnetic field F_{kn} in Equation (18.4) is taken as the given source of the gravitational field, and the Lagrangian $\mathcal{L}_M$ is varied with respect to the gravitational field potentials g_{kn}.

   In order to construct the invariant Lagrangian density of the gravitational field $\mathcal{L}_G$, we note that it is a function of the metric tensor g_{kn} and its first derivatives $\partial_j g_{kn}$ only. On the other hand, by using (9.25) in the form

$$\partial_j g_{kn} = \Gamma_{n,kj} + \Gamma_{k,nj} = g_{nl}\Gamma^l_{kj} + g_{kl}\Gamma^l_{nj}, \tag{18.5}$$

we see that the first derivatives of the metric tensor $\partial_j g_{kn}$ can always be expressed in terms of the suitable metric tensors and Christoffel symbols of the second kind. Thus the invariant Lagrangian density of the gravitational field $\mathcal{L}_G$ can be seen as a function of the metric tensor g_{kn} and the Christoffel symbols of the second kind Γ^j_{kn}. Thus we may write

$$\mathcal{L}_G = \mathcal{L}_G \left(g_{kn}, \Gamma^j_{kn}\right). \tag{18.6}$$

The only nontrivial tensors that can be created from the metric tensor g_{kn} and the Christoffel symbols Γ^j_{kn} are the curvature tensor R^j_{lkn}, defined by (12.14) as

$$R^j_{knl} = \partial_n \Gamma^j_{kl} - \partial_l \Gamma^j_{kn} + \Gamma^p_{kl}\Gamma^j_{pn} - \Gamma^p_{kn}\Gamma^j_{pl}; \tag{18.7}$$

the tensors obtained by contractions of the curvature tensor, e.g., the Ricci tensor R_{kn}, defined by (12.44) as

$$R_{kn} = R^j_{knj} = \partial_n \Gamma^j_{kj} - \partial_j \Gamma^j_{kn} + \Gamma^p_{kj}\Gamma^j_{pn} - \Gamma^p_{kn}\Gamma^j_{pj}; \tag{18.8}$$

and the Ricci scalar R, defined by (12.50) as $R = g^{kn}R_{kn}$.

   Thus, a suitable candidate for the invariant Lagrangian density $\mathcal{L}_G$ is proportional to the Ricci scalar R. In such a case we may choose

$$\mathcal{L}_G = -\frac{c^4}{16\pi G} R, \tag{18.9}$$

where G is the gravitational constant given by (17.43). The factor of proportionality in Equation (18.9) is added to ensure the correct dimensions of the physical quantities. The action integral (18.2) may then be written in the form

$$I_G = -\frac{c^3}{16\pi G} \int_{\Omega} R\sqrt{-g}\, d\Omega. \qquad (18.10)$$

However, from the preceding definitions we note that the Ricci scalar R depends not only on the metric tensor g_{kn} and its first derivatives $\partial_j g_{kn}$, but also on the first-order derivatives of the Christoffel symbols $\partial_l \Gamma^j_{kn}$ and thereby also on the second-order derivatives of the metric tensor. On the other hand, the gravitational field equations must contain up to the second-order derivatives of the metric tensor as the gravitational field potential. As these equations are obtained by the variations of the action (18.2), the Lagrangian density $\mathcal{L}_G$ must not contain the derivatives of the metric tensor of the higher order than the first.

In order to resolve this difficulty, we need to analyze the structure of the Ricci scalar R. If it can be shown that the terms containing the second-order derivatives of the metric tensor do not contribute to the variations of the action integral (18.10), then we could proceed using the action integral (18.10) in the derivation of the gravitational field equations. Since the Ricci scalar is the only available nontrivial scalar invariant created using metric tensor g_{kn} and the Christoffel symbols of the second kind Γ^j_{kn}, it is important to show that the preceding assumption is indeed valid. Let us therefore calculate the quantity $\sqrt{-g}R$ as follows:

$$\sqrt{-g}R = \sqrt{-g}g^{kn}R_{kn} = \sqrt{-g}g^{kn}\partial_n \Gamma^j_{kj}$$
$$- \sqrt{-g}g^{kn}\partial_j \Gamma^j_{kn} + \sqrt{-g}g^{kn}\left(\Gamma^p_{kj}\Gamma^j_{pn} - \Gamma^p_{kn}\Gamma^j_{pj}\right)$$
$$= \partial_n\left(\sqrt{-g}g^{kn}\Gamma^j_{kj}\right) - \partial_j\left(\sqrt{-g}g^{kn}\Gamma^j_{kn}\right) - \partial_n\left(\sqrt{-g}g^{kn}\right)\Gamma^j_{kj}$$
$$+ \partial_j\left(\sqrt{-g}g^{kn}\right)\Gamma^j_{kn} - \sqrt{-g}g^{kn}\left(\Gamma^p_{kn}\Gamma^j_{pj} - \Gamma^p_{kj}\Gamma^j_{pn}\right). \qquad (18.11)$$

Interchanging the dummy indices $n \leftrightarrow j$ in the first term on the right-hand side of Equation (18.11), we obtain

$$\sqrt{-g}R = \partial_j\left(\sqrt{-g}g^{kj}\Gamma^n_{kn} - \sqrt{-g}g^{kn}\Gamma^j_{kn}\right)$$
$$+ \sqrt{-g}\Gamma^j_{kn}\frac{1}{\sqrt{-g}}\partial_j\left(\sqrt{-g}g^{kn}\right) - \sqrt{-g}\Gamma^j_{kj}\frac{1}{\sqrt{-g}}\partial_n\left(\sqrt{-g}g^{kn}\right)$$
$$- \sqrt{-g}g^{kn}\left(\Gamma^p_{kn}\Gamma^j_{pj} - \Gamma^p_{kj}\Gamma^j_{pn}\right). \qquad (18.12)$$

Here we note the result (9.48) in the form

$$g^{jn}\Gamma^k_{jn} = -\frac{1}{\sqrt{-g}}\partial_n\left(\sqrt{-g}g^{kn}\right),$$ (18.13)

and we calculate

$$\frac{1}{\sqrt{-g}}\partial_j\left(\sqrt{-g}g^{kn}\right) = g^{kn}\partial_j\left(\ln\sqrt{-g}\right) + \partial_j g^{kn}.$$ (18.14)

Using the result (9.36) in the form

$$\Gamma^p_{jp} = \partial_j\left(\ln\sqrt{-g}\right),$$ (18.15)

we obtain

$$\frac{1}{\sqrt{-g}}\partial_j\left(\sqrt{-g}g^{kn}\right) = g^{kn}\Gamma^p_{jp} + \partial_j g^{kn}.$$ (18.16)

Using further the result (8.29) applied to the metric tensor g^{kn} and the property that the covariant derivative of the metric tensor is equal to zero, we have

$$D_j g^{kn} = \partial_j g^{kn} + \Gamma^n_{pj}g^{kp} + \Gamma^k_{pj}g^{np} = 0,$$ (18.17)

or

$$\partial_j g^{kn} = -\Gamma^n_{pj}g^{kp} - \Gamma^k_{pj}g^{np}.$$ (18.18)

Substituting (18.18) into (18.16) we obtain

$$\frac{1}{\sqrt{-g}}\partial_j\left(\sqrt{-g}g^{kn}\right) = g^{kn}\Gamma^p_{jp} - \Gamma^n_{pj}g^{kp} - \Gamma^k_{pj}g^{np}.$$ (18.19)

Substituting (18.13) and (18.19) into (18.12) we obtain

$$\sqrt{-g}R = \partial_j\left(\sqrt{-g}g^{kj}\Gamma^n_{kn} - \sqrt{-g}g^{kn}\Gamma^j_{kn}\right)$$
$$+ \sqrt{-g}\Gamma^j_{kn}\left(g^{kn}\Gamma^p_{jp} - \Gamma^n_{pj}g^{kp} - \Gamma^k_{pj}g^{np}\right)$$
$$+ \sqrt{-g}\Gamma^p_{kp}g^{jn}\Gamma^k_{jn} - \sqrt{-g}g^{kn}\left(\Gamma^p_{kn}\Gamma^j_{pj} - \Gamma^p_{kj}\Gamma^j_{pn}\right)$$ (18.20)

or, after regrouping,

$$\sqrt{-g}R = \partial_j\left(\sqrt{-g}g^{kj}\Gamma^n_{kn} - \sqrt{-g}g^{kn}\Gamma^j_{kn}\right)$$
$$+ \sqrt{-g}\left(\Gamma^j_{kn}g^{kn}\Gamma^p_{jp} + \Gamma^p_{kp}g^{jn}\Gamma^k_{jn}\right) - \sqrt{-g}\Gamma^j_{kn}\left(\Gamma^n_{pj}g^{kp} + \Gamma^k_{pj}g^{np}\right)$$
$$- \sqrt{-g}g^{kn}\left(\Gamma^p_{kn}\Gamma^j_{pj} - \Gamma^p_{kj}\Gamma^j_{pn}\right).$$ (18.21)

Interchanging the dummy indices $k \leftrightarrow j$ in the second term on the right-hand side of Equation (18.21) gives

$$\sqrt{-g}R = \partial_j \left(\sqrt{-g} g^{kj} \Gamma_{kn}^n - \sqrt{-g} g^{kn} \Gamma_{kn}^j \right)$$
$$+ \sqrt{-g} g^{kn} \left(\Gamma_{kn}^j \Gamma_{jp}^p + \Gamma_{jp}^p \Gamma_{kn}^j \right) - \sqrt{-g} \left(\Gamma_{kn}^j \Gamma_{pj}^n g^{kp} + \Gamma_{kn}^j \Gamma_{pj}^k g^{np} \right)$$
$$- \sqrt{-g} g^{kn} \left(\Gamma_{kn}^p \Gamma_{pj}^j - \Gamma_{kj}^p \Gamma_{pn}^j \right). \tag{18.22}$$

Now, interchanging all dummy indices in the third term on the right-hand side of Equation (18.22), we obtain

$$\sqrt{-g}R = \partial_j \left(\sqrt{-g} g^{kj} \Gamma_{kn}^n - \sqrt{-g} g^{kn} \Gamma_{kn}^j \right) + 2\sqrt{-g} g^{kn} \Gamma_{kn}^j \Gamma_{jp}^p$$
$$- \sqrt{-g} \left(\Gamma_{kj}^p \Gamma_{np}^j g^{kn} + \Gamma_{jk}^p \Gamma_{np}^j g^{kn} \right) - \sqrt{-g} g^{kn} \left(\Gamma_{kn}^p \Gamma_{pj}^j - \Gamma_{kj}^p \Gamma_{pn}^j \right). \tag{18.23}$$

Using the symmetry of the Christoffel symbols of the second kind with respect to its two lower indices (9.9), the result (18.23) becomes

$$\sqrt{-g}R = \partial_j \left(\sqrt{-g} g^{kj} \Gamma_{kn}^n - \sqrt{-g} g^{kn} \Gamma_{kn}^j \right) + 2\sqrt{-g} g^{kn} \Gamma_{kn}^j \Gamma_{jp}^p$$
$$- 2\sqrt{-g} g^{kn} \Gamma_{kj}^p \Gamma_{pn}^j - \sqrt{-g} g^{kn} \left(\Gamma_{kn}^p \Gamma_{pj}^j - \Gamma_{kj}^p \Gamma_{pn}^j \right) \tag{18.24}$$

or

$$\sqrt{-g}R = \partial_j \left(\sqrt{-g} g^{kj} \Gamma_{kn}^n - \sqrt{-g} g^{kn} \Gamma_{kn}^j \right)$$
$$+ 2\sqrt{-g} g^{kn} \left(\Gamma_{kn}^j \Gamma_{jp}^p - \Gamma_{kj}^p \Gamma_{pn}^j \right) - \sqrt{-g} g^{kn} \left(\Gamma_{kn}^p \Gamma_{pj}^j - \Gamma_{kj}^p \Gamma_{pn}^j \right). \tag{18.25}$$

Now we note that the second term on the right-hand side of Equation (18.24) is twice the third term with a positive sign. Thus we obtain

$$\sqrt{-g}R = \partial_j \left(\sqrt{-g} g^{kj} \Gamma_{kn}^n - \sqrt{-g} g^{kn} \Gamma_{kn}^j \right)$$
$$+ \sqrt{-g} g^{kn} \left(\Gamma_{kn}^p \Gamma_{pj}^j - \Gamma_{kj}^p \Gamma_{pn}^j \right). \tag{18.26}$$

The final result for the quantity $\sqrt{-g}R$ can therefore be written in the form

$$\sqrt{-g}R = \partial_j w^j + \sqrt{-g}\Gamma, \tag{18.27}$$

where we define

$$\Gamma = g^{kn} \left(\Gamma_{kn}^p \Gamma_{pj}^j - \Gamma_{kj}^p \Gamma_{pn}^j \right), \tag{18.28}$$

and

$$w^j = \sqrt{-g} \left(g^{kj} \Gamma_{kn}^n - g^{kn} \Gamma_{kn}^j \right). \tag{18.29}$$

Here we note that the quantity Γ depends only on the metric tensor g_{kn} and its first derivatives $\partial_j g_{kn}$, while the terms containing the second-order derivatives of the metric tensor in the Ricci scalar R are collected into an expression that has the form of the divergence of a vector, i.e., $\partial_j w^j$. Substituting (18.29) into (18.10), we obtain

$$I_G = -\frac{c^3}{16\pi G} \int_\Omega \Gamma \sqrt{-g} \, d\Omega - \frac{c^3}{16\pi G} \int_\Omega \partial_j w^j \, d\Omega. \tag{18.30}$$

According to the Gauss theorem, the second integral on the right-hand side of Equation (18.30) can be transformed into an integral over a hypersurface surrounding the domain Ω over which the integration is carried out in the first integral. Thus the variation of the second integral vanishes because of the variational principle, which requires that the variations of the fields at the limits of the domain Ω are equal to zero. Consequently, we have

$$\delta I_G = -\frac{c^3}{16\pi G} \delta \int_\Omega R \sqrt{-g} \, d\Omega = -\frac{c^3}{16\pi G} \delta \int_\Omega \Gamma \sqrt{-g} \, d\Omega. \tag{18.31}$$

It should be noted that the variation of the action (18.31) is a scalar invariant, although the integral

$$-\frac{c^3}{16\pi G} \int_\Omega \Gamma \sqrt{-g} \, d\Omega \tag{18.32}$$

and the quantity Γ defined by (18.28) are not scalar invariants. Thus we have shown that the terms in the Ricci scalar R containing the second-order derivatives of the metric tensor do not contribute to the variations of the action integral (18.10) and that we may use the action integral (18.10) in the derivation of the gravitational field equations.

18.2 Action for Matter Fields

The variation of the total action (18.1) of the gravitational field in the presence of matter is given by

$$\delta I = \delta I_G + \delta I_M = 0. \tag{18.33}$$

Thus we need to calculate the variations δI_G and δI_M. In this section we consider the variation of the action integral of matter fields I_M, i.e.,

$$\delta I_M = \frac{1}{c} \int_\Omega \delta \left(\sqrt{-g} \mathcal{L}_M \right) d\Omega, \tag{18.34}$$

or

$$\delta I_M = \frac{1}{c} \int_\Omega \left[\frac{\partial \left(\sqrt{-g} \mathcal{L}_M \right)}{\partial g^{kn}} \delta g^{kn} + \frac{\partial \left(\sqrt{-g} \mathcal{L}_M \right)}{\partial \left(\partial_j g^{kn} \right)} \delta \left(\partial_j g^{kn} \right) \right] d\Omega. \tag{18.35}$$

Integrating the second term in the integral (18.35) by parts and dropping the integral over the hypersurface boundary of the domain of integration Ω, following the same steps as in the case of the electromagnetic field, we obtain

$$\delta I_M = \frac{1}{c} \int_\Omega \left\{ \frac{\partial \left(\sqrt{-g} \mathcal{L}_M \right)}{\partial g^{kn}} - \partial_j \left[\frac{\partial \left(\sqrt{-g} \mathcal{L}_M \right)}{\partial \left(\partial_j g^{kn} \right)} \right] \right\} \delta g^{kn} \, d\Omega. \tag{18.36}$$

Let us now define the energy–momentum tensor T_{kn} of the matter fields as follows:

$$\frac{1}{2} \sqrt{-g} T_{kn} = \partial_j \left[\frac{\partial \left(\sqrt{-g} \mathcal{L}_M \right)}{\partial \left(\partial_j g^{kn} \right)} \right] - \frac{\partial \left(\sqrt{-g} \mathcal{L}_M \right)}{\partial g^{kn}}. \tag{18.37}$$

Substituting (18.37) into (18.36) we obtain

$$\delta I_M = -\frac{1}{2c} \int_\Omega T_{kn} \delta g^{kn} \sqrt{-g} \, d\Omega. \tag{18.38}$$

Using here the result $\delta \left(g^{kn} g_{jn} \right) = \delta \delta_j^k = 0$, we may write

$$T_{kn} \delta g^{kn} = g_{kl} T^{lj} g_{nj} \delta g^{kn} = -g_{kl} T^{lj} g^{kn} \delta g_{nj}$$

$$= -\delta_l^n T^{lj} \delta g_{nj} = -T^{nj} \delta g_{nj} = -T^{kn} \delta g_{kn}. \tag{18.39}$$

Substituting (18.39) into (18.38) we obtain

$$\delta I_M = \frac{1}{2c} \int_\Omega T^{kn} \delta g_{kn} \sqrt{-g} \, d\Omega. \tag{18.40}$$

Let us now calculate the energy–momentum tensor for the electromagnetic field with $\mathcal{L}_M$ given by Equation (18.4). In the case of the electromagnetic field the Lagrangian $\mathcal{L}_M$ is a function of the metric tensor g_{kn} only and not of its derivatives $\partial_j g_{kn}$. Thus using (18.37) we obtain the contravariant energy–momentum tensor in the form

$$T^{kn} = -\frac{2}{\sqrt{-g}} \frac{\partial \left(\sqrt{-g} \mathcal{L}_M \right)}{\partial g_{kn}} = -2 \frac{\partial \mathcal{L}_M}{\partial g_{kn}} - \frac{2}{\sqrt{-g}} \frac{\partial \sqrt{-g}}{\partial g_{kn}} \mathcal{L}_M. \tag{18.41}$$

Using the result

$$\frac{2}{\sqrt{-g}} \frac{\partial \sqrt{-g}}{\partial g_{kn}} = \frac{1}{g} \frac{\partial g}{\partial g_{kn}} = \frac{G^{kn}}{g} = g^{kn}, \tag{18.42}$$

we obtain from (18.41)

$$T^{kn} = -2 \frac{\partial \mathcal{L}_M}{\partial g_{kn}} - g^{kn} \mathcal{L}_M. \tag{18.43}$$

Using (18.4), we may write

$$\frac{\partial \mathcal{L}_M}{\partial g_{kn}} = -\frac{1}{4\mu_0} F^{pq} F^{jl} \frac{\partial}{\partial g_{kn}} (g_{pj} g_{ql}) = +\frac{1}{4\mu_0} F^{pq} F^{lj} 2\delta_p^k \delta_j^n g_{ql}$$

$$= +\frac{1}{2\mu_0} F^{kq} F^{ln} g_{ql} = +\frac{1}{2\mu_0} F^{kq} F_q^n. \tag{18.44}$$

Substituting (18.4) and (18.44) into (18.43), we obtain

$$T^{kn} = -\frac{1}{\mu_0} F^{kj} F_j^n + \frac{1}{4\mu_0} g^{kn} F^{jl} F_{jl}. \tag{18.45}$$

The result (18.45) is identical to the result (16.65) obtained by other means in the special theory of relativity with pseudo-Euclidean geometry.

As the next example, let us now consider a continuous distribution of noninteracting particles with a total mass m and a rest energy $\mathcal{E}_0$ in a three-dimensional domain of volume V. If we denote the mass density of the matter distribution by ρ, we may write

$$m = \int_V \rho \, dV, \quad \mathcal{E}_0 = \int_V \rho c^2 \, dV. \tag{18.46}$$

The action integral of the given matter distribution can be written in the form

$$I_K = -mc \int_{S_A}^{S_B} ds = -c \int_V \rho \, dV \int_A^B \sqrt{g_{jl} dx^j dx^l} \tag{18.47}$$

or

$$I_K = -c \int_V \int_{t_A}^{t_B} \frac{\rho}{\sqrt{-g}} \sqrt{g_{jl} \frac{dx^j}{dt} \frac{dx^l}{dt}} \sqrt{-g} dV \, dt. \tag{18.48}$$

Introducing here a system Φ^n,

$$\Phi^n = \frac{\rho}{\sqrt{-g}} \frac{dx^n}{dt}, \tag{18.49}$$

the action integral (18.48) can be written in the form

$$I_K = \frac{1}{c} \int_\Omega \mathcal{L}_{MK} \sqrt{-g}\, d\Omega, \quad \mathcal{L}_{MK} = -c\sqrt{g_{jl}\Phi^j \Phi^l}. \tag{18.50}$$

Since $\mathcal{L}_{MK}$ is the invariant Lagrangian density, we see from (18.50) that the system Φ^n is a contravariant four-vector. The temporal component of the vector Φ^n is proportional to the mass density ρ and is given by

$$\Phi^0 = \frac{\rho c}{\sqrt{-g}}. \tag{18.51}$$

The spatial components of the vector Φ^n are proportional to the flux of mass through a unit of the boundary surface $d\Pi$ of the volume V in the unit of time, i.e.,

$$\Phi^\alpha = \frac{\rho}{\sqrt{-g}} v^\alpha = \frac{d^3 m}{d\Pi_\alpha\, dt}. \tag{18.52}$$

The definitions (18.51) and (18.52) are fully analogous to their respective electromagnetic counterparts (17.59) and (17.60). In analogy with the electromagnetic result (17.63), the mass conservation law gives the following continuity Equation:

$$D_n \Phi^n = \frac{1}{\sqrt{-g}} \frac{\partial}{\partial x^n} \left(\sqrt{-g}\, \Phi^n \right) = 0. \tag{18.53}$$

In the three-dimensional vector notation the continuity Equation (18.53) has the form

$$\frac{\partial \rho}{\partial t} + \text{div}\,(\rho \vec{v}) = 0. \tag{18.54}$$

Integrating over the volume of the three-dimensional domain V and using the Gauss theorem, we obtain the intergal form of the mass conservation law:

$$-\frac{d}{dt} \int_V \rho\, dV = \int_\Pi \rho \vec{v} \cdot d\vec{\Pi}. \tag{18.55}$$

According to the mass conservation law (18.55), the negative increment of mass m within a three-dimensional volume V is equal to the total mass flux through the boundary surface Π of the volume V in the unit of time. Using Equation (18.48), we may write

$$\sqrt{-g}\mathcal{L}_{MK} = -\rho c \frac{ds}{dt} \sqrt{g_{jl} u^j u^l} = -\rho c \frac{ds}{dt}. \tag{18.56}$$

Using (18.41) we obtain the energy–momentum tensor of the given mass distribution as follows:

$$T^{kn} = -\frac{2}{\sqrt{-g}}\frac{\partial\left(\sqrt{-g}\mathcal{L}_{MK}\right)}{\partial g_{kn}} = \frac{\rho c}{\sqrt{-g}}\frac{ds}{dt}u^k u^n. \tag{18.57}$$

In the pseudo-Euclidean case we note that the component T^{00} gives the correct result for the energy density,

$$T^{00} = \frac{\rho c^2}{\sqrt{1 - \frac{v^2}{c^2}}}, \tag{18.58}$$

and the components $T^{0\alpha}$ give the correct result for the momentum density,

$$T^{0\alpha} = \frac{\rho v^\alpha}{\sqrt{1 - \frac{v^2}{c^2}}}. \tag{18.59}$$

For a weak gravitational field at low velocities, the following limits can be used:

$$g_{0\alpha} = g_{\alpha 0} \to 0, \quad g_{\alpha\beta} \to -\delta_{\alpha\beta}, \tag{18.60}$$

and

$$ds \to \sqrt{g_{00}}c\, dt, \quad \sqrt{-g} \to \sqrt{g_{00}}. \tag{18.61}$$

Substituting (18.61) into (18.57), we obtain

$$T^{kn}_{(K)} = \rho c^2 u^k u^n. \tag{18.62}$$

The energy–momentum tensor (18.62) can be considered as the kinetic energy–momentum tensor, which does not include the contribution from the internal energy of the given matter distribution. In order to include the contribution from the internal energy of the matter distribution, we may use the result for the differential of the internal energy $dU = -p\, dV$, where p is the pressure within the matter distribution. Thus we obtain the action integral corresponding to the internal energy in the form

$$I_U = -\int_{t_A}^{t_B} U\, dt = \frac{1}{c}\int_\Omega \mathcal{L}_{MU}\sqrt{-g}\, d\Omega, \tag{18.63}$$

where the Lagrangian density $\mathcal{L}_{MU}$ is given by

$$\mathcal{L}_{MU} = -\frac{dU}{dV} = p. \tag{18.64}$$

As the Lagrangian density $\mathcal{L}_{MU}$ does not include any dependence on the metric tensor g_{kn}, we use the constraint

$$1 - \sqrt{g_{jl}u^j u^l} = 0 \tag{18.65}$$

to create a Lagrangian function $f = f(u^k, \lambda)$ with the constraint

$$f(u^k, \lambda) = p\sqrt{g_{jl}u^j u^l} + \lambda\left(1 - \sqrt{g_{jl}u^j u^l}\right). \tag{18.66}$$

Using the method of Lagrange multipliers, the equations for the parameter λ and the variables u^k are given by

$$\frac{\partial f}{\partial \lambda} = 1 - \sqrt{g_{jl}u^j u^l} = 0 \Rightarrow \sqrt{g_{jl}u^j u^l} = 1 \tag{18.67}$$

and

$$\frac{\partial f}{\partial u^k} = \frac{1}{2}(p - \lambda)\sqrt{g_{jl}u^j u^l} 2g_{kj}u^j = 0 \Rightarrow \lambda = p. \tag{18.68}$$

Using (18.68) we obtain the suitable Lagrangian density for the calculation of the energy–momentum tensor

$$\mathcal{L}_{MU} = p + p\left(1 - \sqrt{g_{jl}u^j u^l}\right) = p. \tag{18.69}$$

The contribution to the energy–momentum tensor from the Lagrangian density (18.69) is obtained using (18.43) as follows:

$$T_{(U)}^{kn} = -2\frac{\partial \mathcal{L}_M}{\partial g_{kn}} - g^{kn}\mathcal{L}_M = p\left(u^k u^n - g^{kn}\right), \tag{18.70}$$

where we used (18.69) to calculate

$$\frac{\partial \mathcal{L}_M}{\partial g_{kn}} = -\frac{1}{2}p\sqrt{g_{jl}u^j u^l}u^k u^n = -\frac{1}{2}pu^k u^n. \tag{18.71}$$

Putting together the results (18.62) and (18.70), we obtain the total energy–momentum tensor of a matter distribution in the form

$$T^{kn} = T_{(K)}^{kn} + T_{(U)}^{kn} = (p + \rho c^2)u^k u^n - g^{kn}p. \tag{18.72}$$

The result (18.72) is the most commonly used form of the energy–momentum tensor of a matter distribution in the general theory of relativity. It should be noted that the quantity T^{00} is always positive, and in most practical calculations the contribution from the pressure terms can be neglected.

18.3 Einstein Field Equations

In this section we consider the variation of the action integral I_G in (18.33), which describes the gravitational field itself, and derive the gravitational field equations. Using the result (18.31) we have

$$\delta I_G = -\frac{c^3}{16\pi G} \delta \int_\Omega R\sqrt{-g}\, d\Omega. \qquad (18.73)$$

In order to calculate the variation (18.73) we need to calculate

$$\delta \left(\sqrt{-g} R \right) = \delta \left(\sqrt{-g} g^{kn} R_{kn} \right)$$

$$= \delta \left(\sqrt{-g} \right) R + \sqrt{-g} \delta g^{kn} R_{kn} + \sqrt{-g} g^{kn} \delta R_{kn}. \qquad (18.74)$$

Using the results (9.35) and (9.40), we may write

$$\delta g = g g^{kn} \delta g_{kn} = -g g_{kn} \delta g^{kn}, \qquad (18.75)$$

and we have

$$\delta \left(\sqrt{-g} \right) = -\frac{1}{2\sqrt{-g}} \delta g = -\frac{1}{2\sqrt{-g}} (-g g_{kn} \delta g^{kn}), \qquad (18.76)$$

or

$$\delta \left(\sqrt{-g} \right) = -\frac{\sqrt{-g}}{2} g_{kn} \delta g^{kn}. \qquad (18.77)$$

Substituting (18.77) into (18.74), we obtain

$$\delta \left(\sqrt{-g} R \right) = \sqrt{-g} \left(R_{kn} - \frac{1}{2} g_{kn} R \right) \delta g^{kn} + \sqrt{-g} g^{kn} \delta R_{kn}. \qquad (18.78)$$

Next we calculate the expression $g^{kn} \delta R_{kn}$ as follows:

$$g^{kn} \delta R_{kn} = g^{kn} \delta \left(\partial_n \Gamma^j_{kj} + \partial_j \Gamma^j_{kn} + \Gamma^p_{kj} \Gamma^j_{pn} - \Gamma^p_{kn} \Gamma^j_{pj} \right)$$

$$= g^{kn} \delta \left(\partial_n \Gamma^j_{kj} - \partial_j \Gamma^j_{kn} \right) - g^{kn} \delta \Gamma^p_{kn} \Gamma^j_{pj} - g^{kn} \Gamma^p_{kn} \delta \Gamma^j_{pj}$$

$$+ g^{kn} \delta \Gamma^p_{kj} \Gamma^j_{pn} + g^{kn} \Gamma^p_{kj} \delta \Gamma^j_{pn} + g^{kn} \Gamma^j_{nj} \delta \Gamma^p_{kp} - g^{kn} \Gamma^j_{nj} \delta \Gamma^p_{kp}$$

$$\qquad (18.79)$$

where we have added the term

$$g^{kn} \Gamma^j_{nj} \delta \Gamma^p_{kp} - g^{kn} \Gamma^j_{nj} \delta \Gamma^p_{kp} = 0 \qquad (18.80)$$

at the end of Equation (18.78). Regrouping the terms, Equation (18.79) becomes

$$g^{kn}\delta R_{kn} = g^{kn}\delta\left(\partial_n\Gamma^j_{kj} - \partial_j\Gamma^j_{kn}\right) + g^{kn}\left(\Gamma^j_{nj}\delta\Gamma^p_{kp} - \Gamma^j_{pj}\delta\Gamma^p_{kn}\right)$$

$$+ g^{kn}\Gamma^p_{kj}\delta\Gamma^j_{pn} + g^{kn}\delta\Gamma^p_{kj}\Gamma^j_{pn} - g^{kn}\Gamma^p_{kn}\delta\Gamma^j_{pj} - g^{kn}\Gamma^j_{nj}\delta\Gamma^p_{kp}$$

$$\text{(18.81)}$$

Renaming the dummy indices, we obtain

$$g^{kn}\delta R_{kn} = g^{kn}\delta\left(\partial_n\Gamma^p_{kp} - \partial_p\Gamma^p_{kn}\right) + g^{kn}\left(\Gamma^l_{nl}\delta\Gamma^p_{kp} - \Gamma^l_{pl}\delta\Gamma^p_{kn}\right)$$

$$+ g^{kn}\Gamma^l_{kp}\delta\Gamma^p_{nl} + g^{kn}\delta\Gamma^l_{kp}\Gamma^p_{nl} - g^{kn}\Gamma^p_{kn}\delta\Gamma^l_{pl} - g^{kn}\Gamma^l_{nl}\delta\Gamma^p_{kp}$$

$$\text{(18.82)}$$

or

$$g^{kn}\delta R_{kn} = g^{kn}\delta\left(\partial_n\Gamma^p_{kp} - \partial_p\Gamma^p_{kn}\right) + g^{kn}\left(\Gamma^l_{nl}\delta\Gamma^p_{kp} - \Gamma^l_{pl}\delta\Gamma^p_{kn}\right)$$

$$+ g^{jn}\Gamma^k_{jp}\delta\Gamma^p_{nk} + g^{kj}\Gamma^n_{jp}\delta\Gamma^p_{kn} - g^{jn}\Gamma^k_{jn}\delta\Gamma^p_{kp} - g^{kj}\Gamma^n_{jn}\delta\Gamma^p_{kp}.$$

$$\text{(18.83)}$$

Using the symmetry of the Christoffel symbols with respect to the lower two indices, we may rewrite (18.83) as

$$g^{kn}\delta R_{kn} = g^{kn}\delta\left(\partial_n\Gamma^p_{kp} - \partial_p\Gamma^p_{kn}\right) + g^{kn}\left(\Gamma^l_{nl}\delta\Gamma^p_{kp} - \Gamma^l_{pl}\delta\Gamma^p_{kn}\right)$$

$$-\left(-g^{jn}\Gamma^k_{jp} - g^{kj}\Gamma^n_{jp}\right)\delta\Gamma^p_{kn} + \left(-g^{jn}\Gamma^k_{jn} - g^{kj}\Gamma^n_{jn}\right)\delta\Gamma^p_{kp}.$$

$$\text{(18.84)}$$

From the results $D_p g^{kn} = 0$ and $D_n g^{kn} = 0$, we have

$$\partial_p g^{kn} = -\Gamma^k_{jp}g^{jn} - \Gamma^n_{jp}g^{kj}$$

$$\partial_n g^{kn} = -\Gamma^k_{jn}g^{jn} - \Gamma^n_{jn}g^{kj}.$$

$$\text{(18.85)}$$

Substituting (18.85) into (18.84) we obtain

$$g^{kn}\delta R_{kn} = g^{kn}\partial_n\left(\delta\Gamma^p_{kp}\right) - g^{kn}\partial_p\left(\delta\Gamma^p_{kn}\right)$$

$$- \delta\Gamma^p_{kn}\partial_p g^{kn} + \delta\Gamma^p_{kp}\partial_n g^{kn} + g^{kn}\left(\Gamma^l_{nl}\delta\Gamma^p_{kp} - \Gamma^l_{pl}\delta\Gamma^p_{kn}\right)$$

$$\text{(18.86)}$$

or

$$g^{kn}\delta R_{kn} = \partial_n \left(g^{kn}\delta\Gamma^p_{kp}\right) - \partial_p \left(g^{kn}\delta\Gamma^p_{kn}\right)$$
$$+ \Gamma^l_{nl}\left(g^{kn}\delta\Gamma^p_{kp}\right) - \Gamma^l_{pl}\left(g^{kn}\delta\Gamma^p_{kn}\right). \tag{18.87}$$

Interchanging the dummy indices $k \leftrightarrow p$ in the first and third term on the right-hand side of Equation (18.87), we obtain

$$g^{kn}\delta R_{kn} = \partial_p \left(g^{kp}\delta\Gamma^n_{kn} - g^{kn}\delta\Gamma^p_{kn}\right)$$
$$+ \Gamma^l_{pl}\left(g^{kp}\delta\Gamma^n_{kn} - g^{kn}\delta\Gamma^p_{kn}\right). \tag{18.88}$$

Introducing here a four-vector ω^p,

$$\omega^p = g^{kp}\delta\Gamma^n_{kn} - g^{kn}\delta\Gamma^p_{kn}, \tag{18.89}$$

and using $\Gamma^l_{pl} = \partial_p \ln \sqrt{-g}$, the result (18.88) becomes

$$g^{kn}\delta R_{kn} = \partial_p \omega^p + \Gamma^l_{pl}\omega^p = \partial_p \omega^p + \frac{1}{\sqrt{-g}}\partial_p \left(\sqrt{-g}\right)\omega^p \tag{18.90}$$

or

$$g^{kn}\delta R_{kn} = \frac{1}{\sqrt{-g}}\partial_p \left(\sqrt{-g}\omega^p\right). \tag{18.91}$$

Substituting (18.91) into the result (18.78) we obtain

$$\delta \left(\sqrt{-g}R\right) = \sqrt{-g}\left(R_{kn} - \frac{1}{2}g_{kn}R\right)\delta g^{kn} + \partial_p \left(\sqrt{-g}\omega^p\right). \tag{18.92}$$

Substituting further (18.92) into (18.73) we obtain

$$\delta I_G = -\frac{c^3}{16\pi G}\int_\Omega \left(R_{kn} - \frac{1}{2}g_{kn}R\right)\delta g^{kn}\sqrt{-g}\,d\Omega$$
$$-\frac{c^3}{16\pi G}\int_\Omega \partial_p \left(\sqrt{-g}\omega^p\right)d\Omega. \tag{18.93}$$

The second integral in (18.93) can, by means of the Gauss theorem, be transformed into an integral of ω^p over the hypersurface surrounding the entire four-dimensional domain Ω. When we vary the action I_G, the variation of the second integral vanishes because of the variational principle, which requires that the variations of the fields at the limits of the domain Ω be equal to zero. Thus the second integral in (18.93) can be dropped and we obtain the final result for the variation of the action integral for the gravitational field in the form

$$\delta I_G = -\frac{c^3}{16\pi G} \int_\Omega \left(R_{kn} - \frac{1}{2}g_{kn}R\right)\delta g^{kn}\sqrt{-g}\,d\Omega. \tag{18.94}$$

Using now the result (18.33), i.e., $-\delta I_G = \delta I_M$ with (18.38) and (18.94), we obtain

$$\frac{c^3}{16\pi G} \int_\Omega \left(R_{kn} - \frac{1}{2}g_{kn}R\right)\delta g^{kn}\sqrt{-g}\,d\Omega$$

$$= -\frac{1}{2c} \int_\Omega T_{kn}\delta g^{kn}\sqrt{-g}\,d\Omega. \tag{18.95}$$

Since the variations of the metric tensor are arbitrary, the result (18.95) gives *Einstein field equations* for the gravitational fields in the form

$$R_{kn} - \frac{1}{2}g_{kn}R = -\frac{8\pi G}{c^4}T_{kn}. \tag{18.96}$$

The field Equations (18.96) with mixed tensors R_n^k and T_n^k are given by

$$R_n^k - \frac{1}{2}\delta_n^k R = -\frac{8\pi G}{c^4}T_n^k. \tag{18.97}$$

Contracting Equation (18.97), we obtain

$$R_n^n - \frac{1}{2}\delta_n^n R = R - 2R = -R = -\frac{8\pi G}{c^4}T. \tag{18.98}$$

Substituting (18.98) into (18.96), we obtain an alternative form for the gravitational field equations:

$$R_{kn} = -\frac{8\pi G}{c^4}\left(T_{kn} - \frac{1}{2}g_{kn}T\right). \tag{18.99}$$

In empty space we have $T_{kn} = 0$, and Equations (18.99) give $R_{kn} = 0$. The result $R_{kn} = 0$, however, does not mean that space-time is flat (i.e., pseudo-Euclidean). The condition for space-time to be pseudo-Euclidean is that the curvature tensor R_{kln}^j is identically equal to zero, i.e., $R_{kln}^j = 0$.

► Chapter 19

Solutions of Field Equations

In the previous chapter we derived the gravitational field equations. The gravitational field equations are nonlinear equations, and they include the self-interaction, i.e., the interaction of the gravitational field with itself. The principle of superposition, which is valid for electromagnetic fields in the special theory of relativity, is therefore not valid for gravitational fields. However, in most practical cases we work with weak gravitational fields for which the field equations can be linearized and the principle of superposition is approximately valid. In this chapter we discuss the nonrelativistic limit of gravitational field equations and the simplest solution of the complete gravitational field equations for the static spherically symmetric field produced by a spherical mass M at rest, known as the Schwarzschild solution. The Schwarzschild solution has played a major role in the early development of the general theory of relativity and is still regarded as a solution of fundamental importance.

19.1 The Newton Law

In order to find the nonrelativistic limit of the gravitational field Equations (18.96) or (18.99), we start with the definition of the Ricci tensor (18.8), i.e.,

$$R_{kn} = \partial_n \Gamma^j_{kj} - \partial_j \Gamma^j_{kn} + \Gamma^p_{kj} \Gamma^j_{pn} - \Gamma^p_{kn} \Gamma^j_{pj}. \tag{19.1}$$

In the nonrelativistic limit in the static gravitational field, with the approximate metric given by (17.37), the only nontrivial Equation (19.1) is the one with $k = n = 0$. Thus we calculate

$$R_{00} = \partial_0 \Gamma^j_{0j} - \partial_j \Gamma^j_{00} + \Gamma^p_{0j}\Gamma^j_{p0} - \Gamma^p_{00}\Gamma^j_{pj} \approx -\partial_\alpha \Gamma^\alpha_{00}. \tag{19.2}$$

For the static gravitational field with $\partial_0 g_{kn} = 0$ we may use the approximation (17.26) with (17.38), such that we have

$$\Gamma^\alpha_{00} = -\frac{1}{2}g^{\alpha\beta}\partial_\beta g_{00} = -\frac{1}{c^2}\partial^\alpha \phi. \tag{19.3}$$

Substituting (19.3) into (19.2), we obtain

$$R_{00} \approx -\partial_\alpha \Gamma^\alpha_{00} = \frac{1}{c^2}\partial_\alpha \partial^\alpha \phi. \tag{19.4}$$

From the result for the energy–momentum tensor of a matter distribution (18.72), in the static nonrelativistic case under consideration, we obtain

$$T_{00} \approx \rho c^2 u_0 u_0 \approx \rho c^2, \quad T \approx \rho c^2 g^{kn} u_k u_n = \rho c^2. \tag{19.5}$$

Thus we obtain

$$T_{00} - \frac{1}{2}g_{00}T = \frac{1}{2}\rho c^2. \tag{19.6}$$

Substituting the results (19.4) and (19.6) into the gravitational field Equation (18.99) with $k = n = 0$, we obtain

$$R_{00} = \frac{1}{c^2}\partial_\alpha \partial^\alpha \phi = -\frac{8\pi G}{c^4}\left(T_{00} - \frac{1}{2}g_{00}T\right) = -\frac{4\pi G}{c^2}\rho, \tag{19.7}$$

or

$$-\partial_\alpha \partial^\alpha \phi = \nabla^2 \phi = 4\pi G\rho. \tag{19.8}$$

Thus we obtain the nonrelativistic gravitational field equation for the gravitational potential ϕ in the form

$$\nabla^2 \phi = 4\pi G\rho. \tag{19.9}$$

The solution of Equation (19.9) is given by

$$\phi(\vec{r}) = -G\int_V \frac{\rho dV}{R}, \tag{19.10}$$

where R is the three-dimensional distance between the position of the sources and the position where the potential is calculated, i.e.,

$$R = \sqrt{|x^\alpha - \bar{x}^\alpha|^2}. \tag{19.11}$$

For a uniform mass distribution over the volume V, we obtain

$$\phi(\vec{r}) = -\frac{G}{R} \int_V \rho dV = -\frac{GM}{R}. \tag{19.12}$$

The result (19.12) agrees with the result (17.41), which leads to the Newton law of gravity (17.44). Thus the gravitational field Equations (18.96) or (18.99) have the correct nonrelativistic limit (19.9), which leads to the Newton law of gravity.

19.2 The Schwarzschild Solution

The Schwarzschild solution is a solution of the complete gravitational field equations for a static spherically symmetric field produced by a spherical mass M at rest. The static condition requires that all the components of the metric tensor g_{kn} be independent of x^0 or time t. Furthermore, we must have $g_{\alpha 0} = g_{0\alpha} = 0$. The spherical symmetry suggests the choice of the spatial coordinates as the three-dimensional spherical coordinates (r, θ, φ). The most general spherically symmetric metric satisfying these conditions can be written in the form

$$ds^2 = e^{2\nu(r)} c^2 dt^2 - e^{2\lambda(r)} dr^2 - r^2 (d\theta^2 + \sin^2 \theta d\varphi^2) \tag{19.13}$$

where $\nu = \nu(r)$ and $\lambda = \lambda(r)$ are yet unspecified functions of the radial coordinate r. In the metric (19.13) we are using the exponential functions in order to secure the right signature of all terms. Thus the covariant metric tensor of a static and spherically symmetric gravitational field is given by

$$[g_{mn}] = \begin{bmatrix} e^{2\nu} & 0 & 0 & 0 \\ 0 & -e^{2\lambda} & 0 & 0 \\ 0 & 0 & -r^2 & 0 \\ 0 & 0 & 0 & -r^2 \sin^2 \theta \end{bmatrix}. \tag{19.14}$$

The contravariant metric tensor is then given by

$$[g_{mn}] = \begin{bmatrix} e^{-2\nu} & 0 & 0 & 0 \\ 0 & -e^{-2\lambda} & 0 & 0 \\ 0 & 0 & -r^{-2} & 0 \\ 0 & 0 & 0 & -r^{-2} \sin^{-2} \theta \end{bmatrix}. \tag{19.15}$$

The Christoffel symbols of the first kind for the metric (19.13) can now be calculated using the definition (9.28), i.e.,

$$\Gamma_{j,kn} = \frac{1}{2}\left(\frac{\partial g_{jk}}{\partial x^n} + \frac{\partial g_{nj}}{\partial x^k} - \frac{\partial g_{kn}}{\partial x^j}\right). \tag{19.16}$$

The results for the Christoffel symbols of the first kind for the metric (19.13) are summarized in the following list:

$$\Gamma_{0,00} = \tfrac{1}{2}(\partial_0 g_{00} + \partial_0 g_{00} - \partial_0 g_{00}) = 0$$

$$\Gamma_{0,01} = \Gamma_{0,10} = \tfrac{1}{2}(\partial_r g_{00} + \partial_0 g_{10} - \partial_0 g_{01}) = v'(r)\exp(2v)$$

$$\Gamma_{0,02} = \Gamma_{0,20} = \tfrac{1}{2}(\partial_\theta g_{00} + \partial_0 g_{20} - \partial_0 g_{02}) = 0$$

$$\Gamma_{0,03} = \Gamma_{0,30} = \tfrac{1}{2}(\partial_\varphi g_{00} + \partial_0 g_{30} - \partial_0 g_{03}) = 0$$

$$\Gamma_{0,11} = \tfrac{1}{2}(\partial_r g_{01} + \partial_r g_{10} - \partial_0 g_{11}) = 0$$

$$\Gamma_{0,12} = \Gamma_{0,21} = \tfrac{1}{2}(\partial_\theta g_{01} + \partial_r g_{20} - \partial_0 g_{12}) = 0$$

$$\Gamma_{0,13} = \Gamma_{0,31} = \tfrac{1}{2}(\partial_\varphi g_{01} + \partial_r g_{30} - \partial_0 g_{13}) = 0$$

$$\Gamma_{0,22} = \tfrac{1}{2}(\partial_\theta g_{02} + \partial_\theta g_{20} - \partial_0 g_{22}) = 0$$

$$\Gamma_{0,23} = \Gamma_{0,32} = \tfrac{1}{2}(\partial_\varphi g_{02} + \partial_\theta g_{30} - \partial_0 g_{23}) = 0$$

$$\Gamma_{0,33} = \tfrac{1}{2}(\partial_\varphi g_{03} + \partial_\varphi g_{30} - \partial_0 g_{33}) = 0$$

$$\Gamma_{1,00} = \tfrac{1}{2}(\partial_0 g_{10} + \partial_0 g_{01} - \partial_r g_{00}) = -v'(r)\exp(2v)$$

$$\Gamma_{1,01} = \Gamma_{1,10} = \tfrac{1}{2}(\partial_r g_{10} + \partial_0 g_{11} - \partial_r g_{01}) = 0$$

$$\Gamma_{1,02} = \Gamma_{1,20} = \tfrac{1}{2}(\partial_\theta g_{10} + \partial_0 g_{21} - \partial_r g_{02}) = 0$$

$$\Gamma_{1,03} = \Gamma_{1,30} = \tfrac{1}{2}(\partial_\varphi g_{10} + \partial_0 g_{31} - \partial_r g_{03}) = 0$$

$$\Gamma_{1,11} = \tfrac{1}{2}(\partial_r g_{11} + \partial_r g_{11} - \partial_r g_{11}) = -\lambda'(r)\exp(2\lambda)$$

$$\Gamma_{1,12} = \Gamma_{1,21} = \tfrac{1}{2}(\partial_\theta g_{11} + \partial_r g_{21} - \partial_r g_{12}) = 0$$

$$\Gamma_{1,13} = \Gamma_{1,31} = \tfrac{1}{2}(\partial_\varphi g_{11} + \partial_r g_{31} - \partial_r g_{13}) = 0$$

$$\Gamma_{1,22} = \tfrac{1}{2}(\partial_\theta g_{12} + \partial_\theta g_{21} - \partial_r g_{22}) = r$$

$$\Gamma_{1,23} = \Gamma_{1,32} = \tfrac{1}{2}(\partial_3 g_{12} + \partial_\theta g_{31} - \partial_r g_{23})$$

$$\Gamma_{1,33} = \tfrac{1}{2}(\partial_\varphi g_{13} + \partial_\varphi g_{31} - \partial_r g_{33}) = r\sin^2\theta$$

$$\Gamma_{2,00} = \tfrac{1}{2}(\partial_0 g_{20} + \partial_0 g_{02} - \partial_\theta g_{00}) = 0$$

$$\Gamma_{2,01} = \Gamma_{2,10} = \tfrac{1}{2}(\partial_r g_{20} + \partial_0 g_{12} - \partial_\theta g_{01}) = 0$$

$$\Gamma_{2,02} = \Gamma_{2,20} = \tfrac{1}{2}(\partial_\theta g_{20} + \partial_0 g_{22} - \partial_\theta g_{02}) = 0$$

$$\Gamma_{2,03} = \Gamma_{2,30} = \tfrac{1}{2}(\partial_\varphi g_{20} + \partial_0 g_{32} - \partial_\theta g_{03}) = 0$$

$$\Gamma_{2,11} = \tfrac{1}{2}(\partial_r g_{21} + \partial_r g_{12} - \partial_\theta g_{11}) = 0$$

$$\Gamma_{2,12} = \Gamma_{2,21} = \tfrac{1}{2}(\partial_\theta g_{21} + \partial_r g_{22} - \partial_\theta g_{12}) = -r$$

$$\Gamma_{2,13} = \Gamma_{2,31} = \tfrac{1}{2}(\partial_\varphi g_{21} + \partial_r g_{32} - \partial_\theta g_{13}) = 0$$

$$\Gamma_{2,22} = \tfrac{1}{2}(\partial_\theta g_{22} + \partial_\theta g_{22} - \partial_\theta g_{22}) = 0$$

$$\Gamma_{2,23} = \Gamma_{2,32} = \tfrac{1}{2}(\partial_\varphi g_{22} + \partial_\theta g_{32} - \partial_\theta g_{23}) = 0$$

$$\Gamma_{2,33} = \tfrac{1}{2}(\partial_\varphi g_{23} + \partial_\varphi g_{32} - \partial_\theta g_{33}) = r^2 \sin\theta \cos\theta$$

$$\Gamma_{3,00} = \tfrac{1}{2}(\partial_0 g_{30} + \partial_0 g_{03} - \partial_\varphi g_{00}) = 0$$

$$\Gamma_{3,01} = \Gamma_{3,10} = \tfrac{1}{2}(\partial_r g_{30} + \partial_0 g_{13} - \partial_\varphi g_{01}) = 0$$

$$\Gamma_{3,02} = \Gamma_{3,20} = \tfrac{1}{2}(\partial_\theta g_{30} + \partial_0 g_{23} - \partial_\varphi g_{02}) = 0$$

$$\Gamma_{3,03} = \Gamma_{3,30} = \tfrac{1}{2}(\partial_\varphi g_{30} + \partial_0 g_{33} - \partial_\varphi g_{03}) = 0$$

$$\Gamma_{3,11} = \tfrac{1}{2}(\partial_r g_{31} + \partial_r g_{13} - \partial_\varphi g_{11}) = 0$$

$$\Gamma_{3,12} = \Gamma_{3,21} = \tfrac{1}{2}(\partial_\theta g_{31} + \partial_r g_{23} - \partial_\varphi g_{12}) = 0$$

$$\Gamma_{3,13} = \Gamma_{3,31} = \tfrac{1}{2}(\partial_\varphi g_{31} + \partial_r g_{33} - \partial_\varphi g_{13}) = -r \sin^2\theta$$

$$\Gamma_{3,22} = \tfrac{1}{2}(\partial_\theta g_{32} + \partial_\theta g_{23} - \partial_\varphi g_{22}) = 0$$

$$\Gamma_{3,23} = \Gamma_{3,32} = \tfrac{1}{2}(\partial_\varphi g_{32} + \partial_\theta g_{33} - \partial_\varphi g_{23}) = -r^2 \sin\theta \cos\theta$$

$$\Gamma_{3,33} = \tfrac{1}{2}(\partial_\varphi g_{33} + \partial_\varphi g_{33} - \partial_\varphi g_{33}) = 0. \tag{19.17}$$

The Christoffel symbols of the second kind for the metric (19.13) can be calculated using the definition (9.29), i.e.,

$$\Gamma^p_{kn} = g^{pj}\Gamma_{j,kn}. \tag{19.18}$$

The results for the Christoffel symbols of the second kind for the metric (19.13) are summarized in the following list:

$$\Gamma^0_{00} = g^{0j}\Gamma_{j,00} = g^{00}\Gamma_{0,00} = 0$$

$$\Gamma^0_{01} = \Gamma^0_{10} = g^{0j}\Gamma_{j,01} = g^{00}\Gamma_{0,01} = v'(r)$$

$$\Gamma^0_{02} = \Gamma^0_{20} = g^{0j}\Gamma_{j,02} = g^{00}\Gamma_{0,02} = 0$$

$$\Gamma^0_{03} = \Gamma^0_{30} = g^{0j}\Gamma_{j,03} = g^{00}\Gamma_{0,03} = 0$$

$$\Gamma_{11}^0 = g^{0j}\Gamma_{j,11} = g^{00}\Gamma_{0,11} = 0$$

$$\Gamma_{12}^0 = \Gamma_{21}^0 = g^{0j}\Gamma_{j,12} = g^{00}\Gamma_{0,12} = 0$$

$$\Gamma_{13}^0 = \Gamma_{31}^0 = g^{0j}\Gamma_{j,13} = g^{00}\Gamma_{0,13} = 0$$

$$\Gamma_{22}^0 = g^{0j}\Gamma_{j,22} = g^{00}\Gamma_{0,22} = 0$$

$$\Gamma_{23}^0 = \Gamma_{32}^0 = g^{0j}\Gamma_{j,23} = g^{00}\Gamma_{0,23} = 0$$

$$\Gamma_{33}^0 = g^{0j}\Gamma_{j,33} = g^{00}\Gamma_{0,33} = 0$$

$$\Gamma_{00}^1 = g^{1j}\Gamma_{j,00} = g^{11}\Gamma_{1,00} = \nu'(r)\exp(2\nu - 2\lambda)$$

$$\Gamma_{01}^1 = \Gamma_{10}^1 = g^{1j}\Gamma_{j,01} = g^{11}\Gamma_{1,01} = 0$$

$$\Gamma_{02}^1 = \Gamma_{20}^1 = g^{1j}\Gamma_{j,02} = g^{11}\Gamma_{1,02} = 0$$

$$\Gamma_{03}^1 = \Gamma_{30}^1 = g^{1j}\Gamma_{j,03} = g^{11}\Gamma_{1,03} = 0$$

$$\Gamma_{11}^1 = g^{1j}\Gamma_{j,11} = g^{11}\Gamma_{1,11} = \lambda'(r)$$

$$\Gamma_{12}^1 = \Gamma_{21}^1 = g^{1j}\Gamma_{j,12} = g^{11}\Gamma_{1,12} = 0$$

$$\Gamma_{13}^1 = \Gamma_{31}^1 = g^{1j}\Gamma_{j,13} = g^{11}\Gamma_{1,13} = 0$$

$$\Gamma_{22}^1 = g^{1j}\Gamma_{j,22} = g^{11}\Gamma_{1,22} = -r\exp(-2\lambda)$$

$$\Gamma_{23}^1 = \Gamma_{32}^1 = g^{1j}\Gamma_{j,23} = g^{11}\Gamma_{1,23} = 0$$

$$\Gamma_{33}^1 = g^{1j}\Gamma_{j,33} = g^{11}\Gamma_{1,33} = -r\sin^2\theta\exp(-2\lambda)$$

$$\Gamma_{00}^2 = g^{2j}\Gamma_{j,00} = g^{22}\Gamma_{2,00} = 0$$

$$\Gamma_{01}^2 = \Gamma_{10}^2 = g^{2j}\Gamma_{j,01} = g^{22}\Gamma_{2,01} = 0$$

$$\Gamma_{02}^2 = \Gamma_{20}^2 = g^{2j}\Gamma_{j,02} = g^{22}\Gamma_{2,02} = 0$$

$$\Gamma_{03}^2 = \Gamma_{30}^2 = g^{2j}\Gamma_{j,03} = g^{22}\Gamma_{2,03} = 0$$

$$\Gamma_{11}^2 = g^{2j}\Gamma_{j,11} = g^{22}\Gamma_{2,11} = 0$$

$$\Gamma_{12}^2 = \Gamma_{21}^2 = g^{2j}\Gamma_{j,12} = g^{22}\Gamma_{2,12} = \frac{1}{r}$$

$$\Gamma_{13}^2 = \Gamma_{31}^2 = g^{2j}\Gamma_{j,13} = g^{22}\Gamma_{2,13} = 0$$

$$\Gamma_{22}^2 = g^{2j}\Gamma_{j,22} = g^{22}\Gamma_{2,22} = 0$$

$$\Gamma_{23}^2 = \Gamma_{32}^2 = g^{2j}\Gamma_{j,23} = g^{22}\Gamma_{2,23} = 0$$

$$\Gamma_{33}^2 = g^{2j}\Gamma_{j,33} = g^{22}\Gamma_{2,33} = -\sin\theta\cos\theta$$

$$\Gamma_{00}^3 = g^{3j}\Gamma_{j,00} = g^{33}\Gamma_{3,00} = 0$$

$$\Gamma^3_{01} = \Gamma^3_{10} = g^{3j}\Gamma_{j,01} = g^{33}\Gamma_{3,01} = 0$$

$$\Gamma^3_{02} = \Gamma^3_{20} = g^{3j}\Gamma_{j,02} = g^{33}\Gamma_{3,02} = 0$$

$$\Gamma^3_{03} = \Gamma^3_{30} = g^{3j}\Gamma_{j,03} = g^{33}\Gamma_{3,03} = 0$$

$$\Gamma^3_{11} = g^{3j}\Gamma_{j,11} = g^{33}\Gamma_{3,11} = 0$$

$$\Gamma^3_{12} = \Gamma^3_{21} = g^{3j}\Gamma_{j,12} = g^{33}\Gamma_{3,12} = 0$$

$$\Gamma^3_{13} = \Gamma^3_{31} = g^{3j}\Gamma_{j,13} = g^{33}\Gamma_{3,13} = \frac{1}{r}$$

$$\Gamma^3_{22} = g^{3j}\Gamma_{j,22} = g^{33}\Gamma_{3,22} = 0$$

$$\Gamma^3_{23} = \Gamma^3_{32} = g^{3j}\Gamma_{j,23} = g^{33}\Gamma_{3,23} = \frac{\cos\theta}{\sin\theta} = \cot\theta$$

$$\Gamma^3_{33} = g^{3j}\Gamma_{j,33} = g^{33}\Gamma_{3,33} = 0. \tag{19.19}$$

From the results listed in (19.19) we can make the following conclusions:

$$\left[\Gamma^\alpha_{0\beta}\right] = \left[\Gamma^\alpha_{\beta 0}\right] = \begin{bmatrix} \Gamma^1_{10} & \Gamma^1_{20} & \Gamma^1_{30} \\ \Gamma^2_{10} & \Gamma^2_{20} & \Gamma^2_{30} \\ \Gamma^3_{10} & \Gamma^3_{20} & \Gamma^3_{30} \end{bmatrix} \equiv 0 \tag{19.20}$$

$$\left[\Gamma^0_{\alpha\beta}\right] = \left[\Gamma^0_{\beta\alpha}\right] = \begin{bmatrix} \Gamma^0_{11} & \Gamma^0_{12} & \Gamma^0_{13} \\ \Gamma^0_{21} & \Gamma^0_{22} & \Gamma^0_{23} \\ \Gamma^0_{31} & \Gamma^0_{32} & \Gamma^0_{33} \end{bmatrix} \equiv 0, \tag{19.21}$$

where $(\alpha, \beta = 1, 2, 3)$. Furthermore, we can calculate

$$\Gamma^j_{0j} = \Gamma^0_{00} + \Gamma^1_{01} + \Gamma^2_{02} + \Gamma^3_{03} = 0$$

$$\Gamma^j_{1j} = \Gamma^0_{10} + \Gamma^1_{11} + \Gamma^2_{12} + \Gamma^3_{13} = \nu' + \lambda' + \frac{2}{r}$$

$$\Gamma^j_{2j} = \Gamma^0_{20} + \Gamma^1_{21} + \Gamma^2_{22} + \Gamma^3_{23} = \cot\theta$$

$$\Gamma^j_{3j} = \Gamma^0_{30} + \Gamma^1_{31} + \Gamma^2_{32} + \Gamma^3_{33} = 0. \tag{19.22}$$

Using the results (19.22) we note that

$$\left[\partial_\beta \Gamma^j_{\alpha j}\right] = \begin{bmatrix} \partial_r \Gamma^j_{1j} & \partial_r \Gamma^j_{2j} & \partial_r \Gamma^j_{3j} \\ \partial_\theta \Gamma^j_{1j} & \partial_\theta \Gamma^j_{2j} & \partial_\theta \Gamma^j_{3j} \\ \partial_\varphi \Gamma^j_{1j} & \partial_\varphi \Gamma^j_{2j} & \partial_\varphi \Gamma^j_{3j} \end{bmatrix} = \begin{bmatrix} \partial_r \Gamma^j_{1j} & 0 & 0 \\ 0 & \partial_\theta \Gamma^j_{2j} & 0 \\ 0 & 0 & 0 \end{bmatrix} \tag{19.23}$$

or

$$\partial_\beta \Gamma^j_{\alpha j} = 0 \quad \text{for} \quad \alpha \neq \beta. \tag{19.24}$$

From the results listed in (19.19) we also have

$$\Gamma^1_{12} = \Gamma^1_{13} = \Gamma^1_{23} = 0 \Rightarrow \partial_r \Gamma^1_{\alpha\beta} = 0 \quad \text{for} \quad \alpha \neq \beta$$

$$\Gamma^2_{12} = \frac{1}{r}, \quad \Gamma^2_{13} = \Gamma^2_{23} = 0 \Rightarrow \partial_\theta \Gamma^2_{\alpha\beta} = 0 \quad \text{for} \quad \alpha \neq \beta$$

$$\Gamma^3_{12} = \Gamma^3_{13} = \Gamma^3_{23} = 0 \Rightarrow \partial_\varphi \Gamma^3_{\alpha\beta} = 0 \quad \text{for} \quad \alpha \neq \beta \tag{19.25}$$

or

$$\partial_\omega \Gamma^\omega_{\alpha\beta} = 0 \quad \text{for} \quad \alpha \neq \beta. \tag{19.26}$$

The Ricci tensor for the metric (19.13) can be calculated using the definition (18.8), i.e.,

$$R_{kn} = \partial_n \Gamma^j_{kj} - \partial_j \Gamma^j_{kn} + \Gamma^p_{kj} \Gamma^j_{pn} - \Gamma^p_{kn} \Gamma^j_{pj}. \tag{19.27}$$

From the result (12.48) we see that the Ricci tensor is a symmetric tensor $R_{kn} = R_{nk}$ and that it has only 10 independent components. The formulae for the components of the Ricci tensor for the metric (19.13) are summarized in the following list:

$$R_{00} = \partial_0 \Gamma^j_{0j} - \partial_j \Gamma^j_{00} + \Gamma^p_{0j} \Gamma^j_{p0} - \Gamma^p_{00} \Gamma^j_{pj}$$

$$R_{01} = R_{10} = \partial_r \Gamma^j_{0j} - \partial_j \Gamma^j_{01} + \Gamma^p_{0j} \Gamma^j_{p1} - \Gamma^p_{01} \Gamma^j_{pj}$$

$$R_{02} = R_{20} = \partial_\theta \Gamma^j_{0j} - \partial_j \Gamma^j_{02} + \Gamma^p_{0j} \Gamma^j_{p2} - \Gamma^p_{02} \Gamma^j_{pj}$$

$$R_{03} = R_{30} = \partial_\varphi \Gamma^j_{0j} - \partial_j \Gamma^j_{03} + \Gamma^p_{0j} \Gamma^j_{p3} - \Gamma^p_{03} \Gamma^j_{pj}$$

$$R_{11} = \partial_r \Gamma^j_{1j} - \partial_j \Gamma^j_{11} + \Gamma^p_{1j} \Gamma^j_{p1} - \Gamma^p_{11} \Gamma^j_{pj}$$

$$R_{12} = R_{21} = \partial_\theta \Gamma^j_{1j} - \partial_j \Gamma^j_{12} + \Gamma^p_{1j} \Gamma^j_{p2} - \Gamma^p_{12} \Gamma^j_{pj}$$

$$R_{13} = R_{31} = \partial_\varphi \Gamma^j_{1j} - \partial_j \Gamma^j_{13} + \Gamma^p_{1j} \Gamma^j_{p3} - \Gamma^p_{13} \Gamma^j_{pj}$$

$$R_{22} = \partial_\theta \Gamma^j_{2j} - \partial_j \Gamma^j_{22} + \Gamma^p_{2j} \Gamma^j_{p2} - \Gamma^p_{22} \Gamma^j_{pj}$$

$$R_{23} = R_{32} = \partial_\varphi \Gamma^j_{2j} - \partial_j \Gamma^j_{23} + \Gamma^p_{2j} \Gamma^j_{p3} - \Gamma^p_{23} \Gamma^j_{pj}$$

$$R_{33} = \partial_\varphi \Gamma^j_{3j} - \partial_j \Gamma^j_{33} + \Gamma^p_{3j} \Gamma^j_{p3} - \Gamma^p_{33} \Gamma^j_{pj}. \tag{19.28}$$

The results for all 10 individual components of the Ricci tensor for the metric (19.13) can be obtained from the list (19.28). However, the calculation process can be facilitated using some of the general results (19.20)–(19.26). Thus for $\alpha = 1, 2, 3$ we may use (19.27) to calculate

$$R_{\alpha 0} = \partial_0 \Gamma^j_{\alpha j} - \partial_j \Gamma^j_{\alpha 0} + \Gamma^p_{\alpha j} \Gamma^j_{p0} - \Gamma^p_{\alpha 0} \Gamma^j_{pj}. \qquad (19.29)$$

Using the condition for the static metric $\partial_0 \Gamma^j_{kn} \equiv 0$, we obtain

$$R_{\alpha 0} = -\partial_\beta \Gamma^\beta_{\alpha 0} - \Gamma^0_{\alpha 0} \Gamma^j_{0j} - \Gamma^\beta_{\alpha 0} \Gamma^j_{\beta j} + \Gamma^p_{\alpha j} \Gamma^j_{p0}. \qquad (19.30)$$

Using $\Gamma^\beta_{\alpha 0} = 0$ and $\Gamma^j_{0j} = 0$ from (19.20) and (19.22), respectively, the result (19.30) becomes

$$R_{\alpha 0} = \Gamma^p_{\alpha j} \Gamma^j_{p0} = \Gamma^0_{\alpha 0} \Gamma^0_{00} + \Gamma^0_{\alpha\beta} \Gamma^\beta_{00} + \Gamma^\beta_{\alpha 0} \Gamma^0_{\beta 0} + \Gamma^\beta_{\alpha\omega} \Gamma^\omega_{\beta 0}. \qquad (19.31)$$

Using $\Gamma^0_{00} = 0$, $\Gamma^\beta_{\alpha 0} = 0$, and $\Gamma^0_{\alpha\beta} = 0$ from (19.19), (19.20), and (19.21), respectively, we see that all four terms on the right-hand side of Equation (19.31) vanish. Thus we obtain

$$R_{\alpha 0} = R_{0\alpha} \equiv 0 \quad (\alpha = 1, 2, 3). \qquad (19.32)$$

Furthermore for $\alpha, \beta = 1, 2, 3$ and $\alpha \neq \beta$, we may use (19.27) to calculate

$$R_{\alpha\beta} = \partial_\beta \Gamma^j_{\alpha j} - \partial_j \Gamma^j_{\alpha\beta} + \Gamma^p_{\alpha j} \Gamma^j_{p\beta} - \Gamma^p_{\alpha\beta} \Gamma^j_{pj}. \qquad (19.33)$$

From (19.24) and (19.26) we see that the first two terms on the right-hand side of Equation (19.33) for $\alpha \neq \beta$ vanish. Using (19.22) we then obtain

$$R_{\alpha\beta} = \Gamma^p_{\alpha j} \Gamma^j_{p\beta} - \Gamma^0_{\alpha\beta} \Gamma^j_{0j} - \Gamma^1_{\alpha\beta} \Gamma^j_{1j} - \Gamma^2_{\alpha\beta} \Gamma^j_{2j}. \qquad (19.34)$$

Using $\Gamma^0_{\alpha\beta} = 0$ and $\Gamma^1_{\alpha\beta} = 0$ for $\alpha \neq \beta$ from (19.21) and (19.25), respectively, we have

$$R_{\alpha\beta} = \Gamma^p_{\alpha j} \Gamma^j_{p\beta} - \delta^1_\alpha \delta^2_\beta \frac{\cot \theta}{r}. \qquad (19.35)$$

Thus the second term on the right-hand side of Equation (19.35) is not equal to zero only for $(\alpha\beta) = (12)$. Let us now calculate the first term on the right-hand side of Equation (19.35), as follows:

$$\Gamma^p_{\alpha j} \Gamma^j_{p\beta} = \Gamma^p_{\alpha 0} \Gamma^0_{p\beta} + \Gamma^p_{\alpha\omega} \Gamma^\omega_{p\beta}, \qquad (19.36)$$

or

$$\Gamma^p_{\alpha j} \Gamma^j_{p\beta} = \Gamma^0_{\alpha 0} \Gamma^0_{0\beta} + \Gamma^\sigma_{\alpha 0} \Gamma^0_{\sigma\beta} + \Gamma^0_{\alpha\omega} \Gamma^\omega_{0\beta} + \Gamma^\sigma_{\alpha\omega} \Gamma^\omega_{\sigma\beta}. \qquad (19.37)$$

From (19.19) we see that the only nonzero term of type $\Gamma^0_{\alpha 0}$ or $\Gamma^0_{0\beta}$ is $\Gamma^0_{01} = v'(r)$. Thus for $\alpha \neq \beta$ the first term on the right-hand side of

Equation (19.37) is zero. Furthermore, using (19.20) and (19.21), we see that the second and third term on the right-hand side of Equation (19.37) are also equal to zero. Thus we have

$$
\begin{aligned}
\Gamma^p_{\alpha j}\Gamma^j_{p\beta} = \Gamma^\sigma_{\alpha\omega}\Gamma^\omega_{\sigma\beta} &= \Gamma^1_{\alpha\omega}\Gamma^\omega_{1\beta} + \Gamma^2_{\alpha\omega}\Gamma^\omega_{2\beta} + \Gamma^3_{\alpha\omega}\Gamma^\omega_{3\beta} \\
&= \Gamma^1_{\alpha 1}\Gamma^1_{1\beta} + \Gamma^1_{\alpha 2}\Gamma^2_{1\beta} + \Gamma^1_{\alpha 3}\Gamma^3_{1\beta} + \Gamma^2_{\alpha 1}\Gamma^1_{2\beta} + \Gamma^2_{\alpha 2}\Gamma^2_{2\beta} + \Gamma^2_{\alpha 3}\Gamma^3_{2\beta} \\
&\quad + \Gamma^3_{\alpha 1}\Gamma^1_{3\beta} + \Gamma^3_{\alpha 2}\Gamma^2_{3\beta} + \Gamma^3_{\alpha 3}\Gamma^3_{3\beta}.
\end{aligned}
\tag{19.38}
$$

Using the result (19.38) we can show by direct calculation that

$$
\Gamma^p_{\alpha j}\Gamma^j_{p\beta} = \delta^1_\alpha \delta^2_\beta \Gamma^3_{13}\Gamma^3_{32} = \delta^1_\alpha \delta^2_\beta \frac{\cot\theta}{r}.
\tag{19.39}
$$

Substituting (19.39) into (19.35) we finally obtain

$$
R_{\alpha\beta} = R_{\beta\alpha} \equiv 0 \quad (\alpha \neq \beta).
\tag{19.40}
$$

From the results (19.32) and (19.40) we conclude that all off-diagonal components of the Ricci tensor are identically equal to zero:

$$
R_{kn} = R_{nk} \equiv 0 \quad (k \neq n).
\tag{19.41}
$$

Thus the only nontrivial components of the Ricci tensor are the diagonal components. Using the static property of the metric $\partial_0 \Gamma^j_{kn} \equiv 0$ and the results (19.20)–(19.26), they can be calculated as

$$
\begin{aligned}
R_{00} &= -\partial_r \Gamma^1_{00} - \Gamma^1_{00}\Gamma^j_{1j} + 2\Gamma^1_{00}\Gamma^0_{10} \\
&= \left(-\nu'' + \nu'\lambda' - \nu'^2 - \frac{2}{r}\nu'\right)\exp(2\nu - 2\lambda) \\
R_{11} &= \partial_r \Gamma^j_{1j} - \partial_r \Gamma^1_{11} - \Gamma^1_{11}\Gamma^j_{1j} + \left(\Gamma^0_{10}\right)^2 + \left(\Gamma^1_{11}\right)^2 \\
&\quad + \left(\Gamma^2_{12}\right)^2 + \left(\Gamma^3_{13}\right)^2 = \nu'' - \nu'\lambda' + \nu'^2 - \frac{2}{r}\lambda' \\
R_{22} &= \partial_\theta \Gamma^j_{2j} - \partial_r \Gamma^1_{22} + 2\Gamma^1_{22}\Gamma^2_{12} + \left(\Gamma^3_{23}\right)^2 - \Gamma^1_{22}\Gamma^j_{1j} \\
&= \left(1 + r\nu' - r\lambda'\right)\exp(-2\lambda) - 1 \\
R_{33} &= -\partial_r \Gamma^1_{33} - \partial_\theta \Gamma^2_{33} - \Gamma^1_{33}\Gamma^j_{1j} - \Gamma^2_{33}\Gamma^j_{2j} + 2\Gamma^1_{33}\Gamma^3_{13} \\
&\quad + 2\Gamma^2_{33}\Gamma^3_{23} = \sin^2\theta\left[\left(1 + r\nu' - r\lambda'\right)\exp(-2\lambda) - 1\right]. \tag{19.42}
\end{aligned}
$$

From the results (19.42) we see that $R_{33} = R_{22}\sin^2\theta$. As $\sin^2\theta$ is in general different from zero, substitution of these two components of the Ricci tensor into the vacuum gravitational field equations $R_{kn} = 0$ gives the same differential equation for v and λ. Thus the vacuum gravitational field equations $R_{kn} = 0$ give only three independent equations for v and λ, i.e.,

$$-v'' + v'\lambda' - v'^2 - \frac{2}{r}v' = 0$$

$$v'' - v'\lambda' + v'^2 - \frac{2}{r}\lambda' = 0$$

$$(1 + rv' - r\lambda')\exp(-2\lambda) = 1. \qquad (19.43)$$

Adding together the first two equations in (19.43) gives

$$-\frac{2}{r}\left(v' + \lambda'\right) = 0 \;\Rightarrow\; \frac{d}{dr}(v + \lambda) = 0. \qquad (19.44)$$

From (19.44) we see that the quantity $v + \lambda$ must be a constant. On the other hand, for large values of radial coordinate r the space-time is approximately flat and both v and λ tend to zero as $r \to \infty$. Thus the constant $v + \lambda$ must be equal to zero, and we have

$$v + \lambda = 0 \;\Rightarrow\; \lambda = -v. \qquad (19.45)$$

Substituting (19.45) into the third of Equations (19.43), we obtain

$$\left(1 + 2rv'\right)\exp(2v) = \frac{d}{dr}\left[r\exp(2v)\right] = 1. \qquad (19.46)$$

Integrating (19.46) we obtain

$$r\exp(2v) = r - r_G, \qquad (19.47)$$

where the integration constant r_G is called the *gravitational radius* of the body. Thus we obtain

$$g_{00} = \exp(2v) = 1 - \frac{r_G}{r}$$

$$g_{11} = \exp(2\lambda) = \left(1 - \frac{r_G}{r}\right)^{-1}. \qquad (19.48)$$

The gravitational radius of the body is defined using the Newtonian limit (17.38) with (19.12) as

$$g_{00} \to 1 + \frac{2\phi}{c^2} = 1 - \frac{2GM}{c^2 r}. \qquad (19.49)$$

Comparing Equation (19.49) with the first of Equations (19.48) we find

$$r_G = \frac{2GM}{c^2}. \qquad (19.50)$$

Thus the final result for the Schwarzschild space-time metric is given by

$$ds^2 = \left(1 - \frac{r_G}{r}\right)c^2 dt^2 - \left(1 - \frac{r_G}{r}\right)^{-1} dr^2 - r^2\left(d\theta^2 + \sin^2\theta d\varphi^2\right).$$

$$(19.51)$$

From the Schwarzschild solution (19.51) we conclude that the empty space outside a spherically symmetric distribution of matter can be described by a static metric. Furthermore, it should be noted that the solution (19.51) is also valid for moving masses as long as the motion preserves the required symmetry, e.g., a centrally symmetric pulsation. We note also that the metric (19.51) depends only on the total mass of the body that is the source of the gravitational field, just as in the case of the Newtonian theory. If we put $dt = 0$ in the metric (19.51) we obtain the three-dimensional space with a line element:

$$dl^2 = \left(1 - \frac{r_G}{r}\right)^{-1} dr^2 + r^2\left(d\theta^2 + \sin^2\theta d\varphi^2\right).$$

$$(19.52)$$

As the Schwarzschild metric tensor g_{kn} is not time dependent, the distances can be defined over a finite portion of space and the integral of the element of the spatial distance dl along a space curve has a definite meaning. In other words it is possible to split the space-time into the space and time with definite meaning. If we can turn the mass M of the source down to zero, the metric (19.51) reduces to the pseudo-Euclidean metric of the flat space-time in the spherical coordinates:

$$ds^2 = c^2 dt^2 - dr^2 - r^2\left(d\theta^2 + \sin^2\theta d\varphi^2\right).$$

$$(19.53)$$

Turning the M on again, we introduce a distortion into both the four-dimensional space-time continuum and the three-dimensional space itself, and neither of them is flat any more. The level of distortion is proportional to the dimensionless quantity r_G/r. In the flat space-time described by the metric (19.53) the radial coordinate r is the measure of the radial distance from the origin of the coordinates. In the curved space-time it is no longer the case and r is just a space coordinate that does not measure the radial distance from the origin. From the line element (19.52) we obtain the square of the infinitesimal radial distance dR for $d\theta = d\varphi = 0$, as follows:

$$dR^2 = \left(1 - \frac{r_G}{r}\right)^{-1} dr^2.$$

$$(19.54)$$

Thus we see that the actual radial distance increment dR is larger than the coordinate differential dr, i.e., $dR > dr$. The distance between two points r_1 and r_2 is then obtained as

$$R_{21} = \int_{r_1}^{r_2} \left(1 - \frac{r_G}{r}\right)^{-1/2} dr$$

$$= \left[\sqrt{r(r - r_G)} + r_G \ln\left(\sqrt{r} + \sqrt{r - r_G}\right)\right]_{r_1}^{r_2} > r_2 - r_1. \quad (19.55)$$

From the result (19.55) it can be shown that for small r_G/r the distance between the two points r_1 and r_2 tends to $r_2 - r_1$. If we take the sphere in space with $r = $ constant, then the three-dimensional space metric (19.52) becomes the metric of a two-dimensional sphere of radius r embedded in the Euclidean space, i.e.,

$$dl^2 = r^2 \left(d\theta^2 + \sin^2\theta d\varphi^2\right). \quad (19.56)$$

The infinitesimal tangential distances are therefore the same as in Euclidean space:

$$dl = r\sqrt{d\theta^2 + \sin^2\theta d\varphi^2}. \quad (19.57)$$

Let us now consider the element of the proper time $d\tau$ given by (17.5) as follows:

$$d\tau = \sqrt{g_{00}}\, dt = \sqrt{1 - \frac{r_G}{r}}\, dt < dt. \quad (19.58)$$

The proper time between any two events occuring at the same point in space is then given by

$$\tau = \int_{t_1}^{t_2} \sqrt{1 - \frac{r_G}{r}}\, dt < t_2 - t_1. \quad (19.59)$$

Thus in the curved space-time near the massive sources of the gravitational field there is a slowing down of time, compared to the time that would be measured in the pseudo-Euclidean limit at the infinite distance from the sources of the gravitational field, i.e., when r tends to infinity.

It should be noted that the Schwarzschild metric (19.51) becomes singular at $r = r_G$, where $g_{00} = 0$ and $g_{11} = -\infty$. However, for most of the observable bodies in the universe the gravitational radius lies well inside them, where the Schwarzschild metric is not applicable anyway. For example, for the Sun the gravitational radius is $r_G = 2.9$ km and for the Earth the gravitational radius is $r_G = 0.88$ cm. Furthermore the singularity of the Schwarzschild metric can be shown to be more a consequence of the choice of the space-time coordinates than of the space-time itself.

Nevertheless for a few bodies in the universe that are actually smaller than their gravitational radius, some physically interesting things do happen at the boundary $r = r_G$. For example, matter and energy may fall into the region where $r < r_G$ but neither matter nor energy (including the light signals) can escape from the region where $r < r_G$. Such a region is called a *black hole* and will be discussed in the next chapter.

► Chapter 20

Applications of the Schwarzschild Metric

In the previous chapter we derived the static Schwarzschild solution of the gravitational field equations for a static spherically symmetric field produced by a spherical mass M at rest. In this chapter we discuss two applications of the Schwarzschild solution to explain two physical phenomena that cannot be explained within the framework of the classical Newtonian theory of gravitation.

20.1 The Perihelion Advance

According to the Newtonian theory of gravitation, the orbit of a planet around the Sun is a closed ellipse with the Sun at one of the two foci. Thus the point, which is called the *perihelion* and where the planet is closest to the Sun, is fixed. However, experimental evidence shows that the perihelion of the planets is not fixed, but gradually rotates around the Sun. This rotation, although very slow, is cumulative and can be measured over a long period of time. In classical mechanics the Lagrangian of a planet in the central field of the Sun is given by

$$L = \mathcal{E}_k - m\phi(r) = \frac{mv^2}{2} + \frac{GMm}{r}. \tag{20.1}$$

The conserved energy of the planet is

$$E = \frac{\partial L}{\partial v^\alpha} v^\alpha - L = \frac{mv^2}{2} - \frac{GMm}{r} = \text{Constant.} \tag{20.2}$$

From (20.1) we may write in the spherical coordinates

$$L = m \left[\frac{1}{2} (\dot{r}^2 + r^2\dot{\theta}^2 + r^2 \sin^2\theta \dot{\varphi}^2) + \frac{GM}{r} \right], \tag{20.3}$$

where the dots denote the time differentiation of coordinates. Since the classical motion of the planet in the central field of the Sun is confined to the plane of the orbit, which we take to be the equatorial plane of the spherical coordinates, we have $\theta = \pi/2$ and $\dot{\theta} = 0$. From (20.3) we then obtain

$$L = m \left[\frac{1}{2} (\dot{r}^2 + r^2\dot{\varphi}^2) + \frac{GM}{r} \right]. \tag{20.4}$$

The Lagrangian equation with respect to the angular variable φ is given by

$$\frac{d}{dt} \left(\frac{\partial L}{\partial \dot{\varphi}} \right) = \frac{\partial L}{\partial \varphi}. \tag{20.5}$$

Substituting (20.4) into (20.5) we obtain

$$\frac{d}{dt}(mr^2\dot{\varphi}) = 0, \tag{20.6}$$

or

$$mr^2\dot{\varphi} = h = \text{Constant} \Rightarrow r^2\dot{\varphi} = \frac{h}{m}, \tag{20.7}$$

where h is the conserved angular momentum of the planet. Substituting (20.7) into (20.2) in the polar coordinates, we obtain

$$E = \frac{m}{2} \left[\dot{r}^2 + \frac{h^2}{m^2 r^2} - \frac{2GM}{r} \right]. \tag{20.8}$$

From (20.8) we may write

$$\dot{r}^2 = \frac{2E}{m} - \frac{h^2}{m^2 r^2} + \frac{2GM}{r}. \tag{20.9}$$

Dividing now $\dot{r}^2$ from (20.9) by $r^4\dot{\varphi}^2$ from (20.7), we obtain

$$\frac{1}{r^4} \left(\frac{dr}{d\varphi} \right)^2 = \frac{m^2}{h^2} \left(\frac{2E}{m} - \frac{h^2}{m^2 r^2} + \frac{2GM}{r} \right), \tag{20.10}$$

or

$$\left[\frac{d}{d\varphi} \left(\frac{1}{r} \right) \right]^2 = \frac{2mE}{h^2} - \left(\frac{1}{r} \right)^2 + \frac{2GMm^2}{h^2} \left(\frac{1}{r} \right).$$ (20.11)

Introducing here a new variable $u = 1/r$, we obtain

$$\left(\frac{du}{d\varphi} \right)^2 + u^2 = \frac{2mE}{h^2} + \frac{2GMm^2}{h^2} u,$$ (20.12)

or

$$\left(\frac{du}{d\varphi} \right)^2 = \frac{2mE}{h^2} + \frac{G^2 M^2 m^4}{h^4} - \left(\frac{GMm^2}{h^2} - u \right)^2.$$ (20.13)

Introducing here the notation

$$\frac{e^2}{p^2} = \frac{2mE}{h^2} + \frac{G^2 M^2 m^4}{h^4}, \quad \frac{1}{p} = \frac{GMm^2}{h^2},$$ (20.14)

where

$$e = 1 + \frac{2Eh^2}{G^2 M^2 m^3}, \quad p = \frac{h^2}{GMm^2},$$ (20.15)

we may rewrite (20.13) as follows:

$$\left(\frac{du}{d\varphi} \right)^2 = \frac{e^2}{p^2} - \left(\frac{1}{p} - u \right)^2.$$ (20.16)

The well-known solution of the differential Equation (20.16) is given by

$$u = \frac{1 + e \cos \varphi}{p}.$$ (20.17)

The validity of (20.17) is easily confirmed by direct substitution. Using (20.17) we can calculate

$$\left(\frac{1}{p} - u \right)^2 = \frac{e^2}{p^2} \cos^2 \varphi, \quad \left(\frac{du}{d\varphi} \right)^2 = \frac{e^2}{p^2} \sin^2 \varphi.$$ (20.18)

Substituting (20.18) into (20.16) and using $\cos^2 \varphi + \sin^2 \varphi = 1$ we obtain the identity. Thus (20.17) is a correct solution of the differential Equation (20.16). Using here $u = 1/r$ we finally obtain

$$r = \frac{p}{1 + e \cos \varphi}.$$ (20.19)

This is the equation of an ellipse in polar coordinates. The length of the major axis of this ellipse, denoted by $2a$, is given by

$$2a = r(\varphi = 0) + r(\varphi = \pi) = \frac{p}{1+e} + \frac{p}{1-e} = \frac{2p}{1-e^2}. \qquad (20.20)$$

Thus we may express the parameter p as

$$p = a(1 - e^2). \qquad (20.21)$$

Thus according to the Newtonian theory of gravitation the orbit of a planet around the Sun is indeed a closed ellipse with the Sun at one of the two foci, and the perihelion is fixed. Let us now consider the motion of the planet in the Schwarzschild space-time within the framework of the general theory of relativity. The equations of motion are the geodesic Equations (11.25), i.e.,

$$\frac{d^2 x^n}{ds^2} + \Gamma^n_{lk} \frac{dx^l}{ds} \frac{dx^k}{ds} = 0 \quad (k,l,n = 0,1,2,3). \qquad (20.22)$$

Using the nonzero Christoffel symbols of the second kind in the Schwarzschild metric (19.19), we obtain the four equations of motion:

$$\frac{d^2 t}{ds^2} + 2\Gamma^0_{01} \frac{dt}{ds} \frac{dr}{ds} = 0$$

$$\frac{d^2 r}{ds^2} + c^2 \Gamma^1_{00} \left(\frac{dt}{ds}\right)^2 + \Gamma^1_{11} \left(\frac{dr}{ds}\right)^2 + \Gamma^1_{22} \left(\frac{d\theta}{ds}\right)^2 + \Gamma^1_{33} \left(\frac{d\varphi}{ds}\right)^2 = 0$$

$$\frac{d^2 \theta}{ds^2} + 2\Gamma^2_{12} \frac{dr}{ds} \frac{d\theta}{ds} + \Gamma^2_{33} \left(\frac{d\varphi}{ds}\right)^2 = 0$$

$$\frac{d^2 \varphi}{ds^2} + 2\Gamma^3_{13} \frac{dr}{ds} \frac{d\varphi}{ds} + 2\Gamma^3_{23} \frac{d\theta}{ds} \frac{d\varphi}{ds} = 0. \qquad (20.23)$$

In Equations (20.23) the factor of 2 appears before each of the off-diagonal terms, as they appear twice in the sum because of the symmetry of the Christoffel symbols of the second kind with respect to its two lower indices. The Schwarzschild metric is a spherically symmetric metric and the motion of the planets around the Sun is still confined to the plane of the orbit, which we may take to be the equatorial plane of the spherical coordinates. Thus we still have $\theta = \pi/2$ and $\dot{\theta} = 0$. The third of the equations of motion (20.23) for the angular variable θ becomes, therefore, trivial, and the other three equations are reduced to:

$$\frac{d^2t}{ds^2} + 2\Gamma_{01}^0 \frac{dt}{ds}\frac{dr}{ds} = 0$$

$$\frac{d^2r}{ds^2} + c^2\Gamma_{00}^1 \left(\frac{dt}{ds}\right)^2 + \Gamma_{11}^1 \left(\frac{dr}{ds}\right)^2 + \Gamma_{33}^1 \left(\frac{d\varphi}{ds}\right)^2 = 0$$

$$\frac{d^2\varphi}{ds^2} + 2\Gamma_{13}^3 \frac{dr}{ds}\frac{d\varphi}{ds} = 0. \tag{20.24}$$

Substituting the actual values for the Christoffel symbols of the second kind from (19.19) in the first and the third of Equations (20.24), we obtain

$$\frac{d^2t}{ds^2} + 2v'(r)\frac{dr}{ds}\frac{dt}{ds} = 0$$

$$\frac{d^2\varphi}{ds^2} + \frac{2}{r}\frac{dr}{ds}\frac{d\varphi}{ds} = 0, \tag{20.25}$$

or

$$\exp\left(-2v\right)\frac{d}{ds}\left(\exp\left(2v\right)\frac{dt}{ds}\right) = 0$$

$$\frac{1}{r^2}\frac{d}{ds}\left(r^2\frac{d\varphi}{ds}\right) = 0. \tag{20.26}$$

Thus we obtain two constants of motion:

$$\exp 2v\frac{dt}{ds} = \left(1 - \frac{r_G}{r}\right)\frac{dt}{ds} = \frac{1}{c}\sqrt{1 + \frac{2E}{mc^2}} = \text{Constant}$$

$$r^2\frac{d\varphi}{ds} = \frac{h}{mc} = \text{Constant}. \tag{20.27}$$

The form of the constants on the right-hand side of Equations (20.27) is chosen in such a way to secure the correct classical analogy. From Equations (20.27) we may write

$$c^2\left(\frac{dt}{ds}\right)^2 = \left(1 - \frac{r_G}{r}\right)^{-2}\left(1 + \frac{2E}{mc^2}\right)$$

$$\left(r^2\frac{d\varphi}{ds}\right)^2 = \frac{h^2}{m^2c^2}. \tag{20.28}$$

Now, using the result for the Schwarzschild metric (19.51) with $\theta = \pi/2$ and $\dot{\theta} = 0$, we have

$$ds^2 = \left(1 - \frac{r_G}{r}\right)c^2 dt^2 - \left(1 - \frac{r_G}{r}\right)^{-1} dr^2 - r^2 d\varphi^2 \qquad (20.29)$$

or

$$1 = \left(1 - \frac{r_G}{r}\right)c^2 \left(\frac{dt}{ds}\right)^2 - \left(1 - \frac{r_G}{r}\right)^{-1} \left(\frac{dr}{ds}\right)^2 - \frac{1}{r^2}\left(r^2 \frac{d\varphi}{ds}\right)^2. \qquad (20.30)$$

Substituting (20.28) into (20.30) we obtain

$$1 = \left(1 - \frac{r_G}{r}\right)^{-1}\left(1 + \frac{2E}{mc^2}\right)$$
$$- \left(1 - \frac{r_G}{r}\right)^{-1}\left(\frac{dr}{ds}\right)^2 - \frac{h^2}{m^2 c^2 r^2}, \qquad (20.31)$$

or

$$1 - \frac{r_G}{r} = 1 + \frac{2E}{mc^2} - \left(\frac{dr}{ds}\right)^2 - \frac{h^2}{m^2 c^2 r^2}\left(1 - \frac{r_G}{r}\right). \qquad (20.32)$$

Using (20.27) we may calculate

$$\frac{dr}{ds} = \frac{1}{r^2}\frac{dr}{d\varphi}r^2\frac{d\varphi}{ds} = -\frac{h}{mc}\frac{d}{d\varphi}\left(\frac{1}{r}\right). \qquad (20.33)$$

Substituting (20.33) into (20.32) we obtain

$$\frac{h^2}{m^2 c^2}\left[\frac{d}{d\varphi}\left(\frac{1}{r}\right)\right]^2 + \frac{h^2}{m^2 c^2 r^2} = \frac{2E}{mc^2} + \frac{r_G}{r} + \frac{h^2 r_G}{m^2 c^2 r^3}. \qquad (20.34)$$

Using here the result (19.50) for r_G and introducing a new variable $u = 1/r$, we obtain

$$\frac{h^2}{m^2 c^2}\left[\left(\frac{du}{d\varphi}\right)^2 + u^2\right] = \frac{2E}{mc^2} + \frac{2GM}{c^2}u + \frac{h^2 r_G}{m^2 c^2}u^3 \qquad (20.35)$$

or

$$\left(\frac{du}{d\varphi}\right)^2 + u^2 = \frac{2mE}{h^2} + \frac{2GMm^2}{h^2}u + r_G u^3. \qquad (20.36)$$

Using again the result (19.50) for r_G, we finally obtain

$$\left(\frac{du}{d\varphi}\right)^2 + u^2 = \frac{2mE}{h^2} + \frac{2GMm^2}{h^2}u + \frac{2GM}{c^2}u^3. \qquad (20.37)$$

Comparing Equation (20.37) with the Newtonian Equation (20.12) we see that the only difference is an additional nonlinear term proportional to u^3 that vanishes for $c \to \infty$. This term can therefore be considered as a small perturbation of the Newtonian Equation (20.12) that introduces the general-relativistic corrections to the Newtonian results. Differentiating both sides of Equation (20.37) with respect to the angular variable φ, we obtain

$$2\frac{du}{d\varphi}\left(\frac{d^2u}{d\varphi^2} + u\right) = 2\frac{du}{d\varphi}\left(\frac{GMm^2}{h^2} + \frac{3GM}{c^2}u^2\right). \qquad (20.38)$$

If we disregard the solution $du/d\varphi = 0$, which represents the circular orbit with $r = $ constant, we obtain the nonlinear differential equation for u in the form

$$\frac{d^2u}{d\varphi^2} + u = \frac{GMm^2}{h^2} + \frac{3GM}{c^2}u^2. \qquad (20.39)$$

The solution to the corresponding Newtonian linear equation,

$$\frac{d^2u^{(0)}}{d\varphi^2} + u^{(0)} = \frac{GMm^2}{h^2}, \qquad (20.40)$$

is given by Equation (20.17), i.e.,

$$u^{(0)} = \frac{1 + e\cos\varphi}{p}. \qquad (20.41)$$

Using the perturbation method, we may construct an approximate solution of the nonlinear Equation (20.39) in the form

$$u = u^{(0)} + u^{(1)}, \quad u^{(1)} \ll u^{(0)}. \qquad (20.42)$$

Using (20.42) with (20.41) we may write

$$\frac{d^2u}{d\varphi^2} = \frac{d^2u^{(0)}}{d\varphi^2} + \frac{d^2u^{(1)}}{d\varphi^2} = \frac{d^2u^{(1)}}{d\varphi^2} - \frac{e}{p}\cos\varphi, \qquad (20.43)$$

and

$$u = u^{(0)} + u^{(1)} = u^{(1)} + \frac{e}{p}\cos\varphi + \frac{1}{p}. \qquad (20.44)$$

Combining the results (20.43) and (20.44) and using the definition (20.14) we obtain

$$\frac{d^2u}{d\varphi^2} + u = \frac{d^2u^{(1)}}{d\varphi^2} + u^{(1)} + \frac{GMm^2}{h^2}. \tag{20.45}$$

Substituting the result (20.45) into Equation (20.39) and putting $u = u^{(0)}$ in the nonlinear term, we obtain the equation for the perturbation $u^{(1)}$ in the form

$$\frac{d^2u^{(1)}}{d\varphi^2} + u^{(1)} = \frac{3GM}{c^2}(u^{(0)})^2. \tag{20.46}$$

Substituting the result (20.41) into (20.46) we obtain

$$\frac{d^2u^{(1)}}{d\varphi^2} + u^{(1)} = \frac{3GM}{c^2p^2}(1 + e\cos\varphi)^2, \tag{20.47}$$

or

$$\frac{d^2u^{(1)}}{d\varphi^2} + u^{(1)} = \frac{3GM}{c^2p^2} + \frac{6GM}{c^2p^2}e\cos\varphi + \frac{3GM}{c^2p^2}e^2\cos^2\varphi. \tag{20.48}$$

Considering orbits with small eccentricity e the term of the order e^2 can be neglected. The contribution from the constant term is negligible as well. The only term that produces an observable effect is the one proportional to $\cos\varphi$ with the contribution which increases continuously after each revolution. Dropping the constant term and the term proportional to e^2, we obtain

$$\frac{d^2u^{(1)}}{d\varphi^2} + u^{(1)} = \frac{6GM}{c^2p^2}e\cos\varphi. \tag{20.49}$$

The solution of differential Equation (20.49) is given by

$$u^{(1)} = \frac{3GM}{c^2p^2}e\,\varphi\sin\varphi, \tag{20.50}$$

which is easily shown by direct substitution into (20.49). Substituting (20.41) and (20.50) into (20.42), we obtain

$$u = \frac{1 + e\cos\varphi}{p} + \frac{3GM}{c^2p}\frac{e}{p}\varphi\sin\varphi. \tag{20.51}$$

Let us now introduce the increment $\Delta\varphi$ as follows:

$$\Delta\varphi = \frac{3GM}{c^2p}\varphi = \frac{3GM}{c^2a(1 - e^2)}\varphi, \tag{20.52}$$

where we have used the result (20.21) for the parameter p. Thus we may rewrite (20.51) as follows:

$$u = \frac{1 + e\cos\varphi + e\,\Delta\varphi\sin\varphi}{p}. \tag{20.53}$$

Using the trigonometric formula

$$\cos(\varphi - \Delta\varphi) = \cos\varphi\cos\Delta\varphi + \sin\varphi\sin\Delta\varphi \tag{20.54}$$

and the approximations for the trigonometric functions of a small angle $\Delta\varphi$, given by

$$\cos\Delta\varphi \approx 1, \quad \sin\Delta\varphi \approx \Delta\varphi, \tag{20.55}$$

we obtain

$$\cos\varphi + \Delta\varphi\sin\Delta\varphi \approx \cos(\varphi - \Delta\varphi). \tag{20.56}$$

Substituting (20.56) into (20.53), we finally obtain

$$u = \frac{1 + e\cos(\varphi - \Delta\varphi)}{p}. \tag{20.57}$$

From Equation (20.57) we see that while a planet moves through an angle φ, the perihelion advances by a fraction of the revolution angle equal to

$$\frac{\Delta\varphi}{\varphi} = \frac{3GM}{c^2 a(1 - e^2)}. \tag{20.58}$$

For a complete revolution ($\varphi = 2\pi$) the perihelion advances by an angle

$$\Delta\varphi = \frac{6\pi GM}{c^2 a(1 - e^2)}. \tag{20.59}$$

For the planet Mercury the theoretical result for the relativistic perihelion shift is equal to $\Delta\varphi = 43.03$ seconds of arc per century, while the observed perihelion shift is equal to $\Delta\varphi = 43.11 \pm 0.45$ seconds of arc per century. The perihelion shift of the planet Mercury could not be explained in the Newtonian theory of gravitation. This remarkable agreement of the theoretical result from the general theory of relativity with the observational data was the first major experimental confirmation of the theory.

20.2 Black Holes

In the previous chapter we mentioned the singularity of the Schwarzschild metric (19.51) at the gravitational radius of the spherical body $r = r_G$ and the possibility of the existence of bodies in the universe that are actually smaller than their gravitational radius r_G. The region where $r < r_G$ around such bodies is called a *black hole*. A number of physically interesting phenomena occur at the boundary $r = r_G$. For example, we have argued that matter and energy may fall into the region $r < r_G$ of a black hole, but neither matter nor energy (including the light signals) can escape from that region. In order to prove this assertion, let us consider a particle falling radially into a black hole with a radial velocity $u^1 = dr/ds$. As the particle is falling radially, we have $u^2 = u^3 = 0$. The motion of the particle is described by geodesic Equations (11.25), i.e.,

$$\frac{du^n}{ds} + \Gamma^n_{lk} u^l u^k = 0. \tag{20.60}$$

Using $u^2 = u^3 = 0$ and the results for the Christoffel symbols of the second kind (19.19), we may write the temporal equation of motion as follows

$$\frac{du^0}{ds} = -\Gamma^0_{lk} u^l u^k = -2\Gamma^0_{10} u^1 u^0, \tag{20.61}$$

or

$$\frac{du^0}{ds} = -2\frac{dv}{dr}\frac{dr}{ds} u^0 = -2\frac{dv}{ds} u^0, \tag{20.62}$$

where the factor of 2 appears because the term proportional to $\Gamma^0_{10} = \Gamma^0_{01}$ appear twice in the sum. Thus we obtain

$$\frac{du^0}{ds} + 2\frac{dv}{ds} u^0 = \exp(-2v)\frac{d}{ds}\left(\exp(2v)u^0\right) = 0. \tag{20.63}$$

Equation (20.63) can be integrated to give

$$\exp(2v)u^0 = g_{00}u^0 = K = \text{Constant}, \tag{20.64}$$

where K is the integration constant that is equal to the value of g_{00} at the point where the particle starts to fall toward the black hole. Using the identity $g_{kn}u^k u^n \equiv 1$, we may also write

$$1 = g_{kn}u^k u^n = g_{00}(u^0)^2 + g_{11}(u^1)^2. \tag{20.65}$$

Multiplying by g_{00} and using (20.64) as well as $g_{00}g_{11} = -1$, we obtain

$$g_{00} = (g_{00})^2(u^0)^2 + g_{00}g_{11}(u^1)^2 = K^2 - (u^1)^2, \tag{20.66}$$

or

$$(u^1)^2 = K^2 - g_{00} = K^2 - 1 + \frac{r_G}{r}. \tag{20.67}$$

For the radially falling body we have $u^1 < 0$, and we may write

$$u^1 = -\left(K^2 - 1 + \frac{r_G}{r}\right)^{1/2}. \tag{20.68}$$

Using (20.64) and (20.68) we may calculate the ratio dt/dr as follows:

$$\frac{dt}{dr} = \frac{u^0}{cu^1} = -\frac{K}{cg_{00}}\left(K^2 - 1 + \frac{r_G}{r}\right)^{-1/2}, \tag{20.69}$$

or

$$\frac{dt}{dr} = -\frac{K}{c}\left(1 - \frac{r_G}{r}\right)^{-1}\left(K^2 - 1 + \frac{r_G}{r}\right)^{-1/2}. \tag{20.70}$$

If we now assume that the particle falling radially into a black hole is close to the gravitational radius r_G, then we may write $r = r_G + \epsilon$ with $\epsilon \ll r_G$. Equation (20.70) then becomes

$$\frac{dt}{dr} = -\frac{K}{c}\left[1 - \left(1 + \frac{\epsilon}{r_G}\right)^{-1}\right]^{-1}$$

$$\times \left[K^2 - 1 + \left(1 + \frac{\epsilon}{r_G}\right)^{-1}\right]^{-1/2}. \tag{20.71}$$

Using the approximation

$$\left(1 + \frac{\epsilon}{r_G}\right)^{-1} \approx 1 - \frac{\epsilon}{r_G}, \tag{20.72}$$

in Equation (20.71), we obtain

$$\frac{dt}{dr} \approx -\frac{Kr_G}{c\epsilon}\left[K^2 - \frac{\epsilon}{r_G}\right]^{-1/2} \approx -\frac{r_G}{c\epsilon}, \tag{20.73}$$

or

$$\frac{dt}{dr} \approx -\frac{r_G}{c\,(r - r_G)}. \tag{20.74}$$

Integrating (20.74) we obtain

$$t \approx -r_G \ln\,(r - r_G) + \text{Constant}. \tag{20.75}$$

From Equation (20.75) we note that as $r \to r_G$ we have $t \to \infty$. Let us now consider an observer traveling with the particle. The proper time measured by the observer in its own rest frame is given by (17.5), i.e.,

$$d\tau = \frac{1}{c}ds = \sqrt{g_{00}}dt. \tag{20.76}$$

Thus we may write

$$d\tau = \frac{1}{c}ds = \frac{1}{c}\frac{ds}{dr}dr = \frac{1}{cu^1}dr, \tag{20.77}$$

or

$$d\tau = -\frac{dr}{c}\left(K^2 - 1 + \frac{r_G}{r}\right)^{-1/2}. \tag{20.78}$$

If we again assume that the observer traveling radially into a black hole is very close to the gravitational radius r_G, then we have $r = r_G + \epsilon$ with $dr = d\epsilon$. Equation (20.78) then becomes

$$d\tau = -\frac{d\epsilon}{c}\left(K^2 - \frac{\epsilon}{r_G + \epsilon}\right)^{-1/2}, \tag{20.79}$$

or

$$d\tau \approx -\frac{d\epsilon}{c}\left(K^2 - \frac{\epsilon}{r_G}\right)^{-1/2}. \tag{20.80}$$

Integrating from the starting point ϵ to the point $\epsilon = 0$ where the observer reaches the gravitational radius r_G, we obtain

$$
\begin{aligned}
\tau &\approx -\frac{1}{c}\int_{\epsilon}^{0}\left(K^2 - \frac{\epsilon}{r_G}\right)^{-1/2}d\epsilon \\
&= \frac{2r_G K}{c}\left(1 - \frac{\epsilon}{K^2 r_G}\right)^{1/2}\Bigg|_{\epsilon}^{0} \\
&= \frac{2r_G K}{c}\left[1 - \sqrt{1 - \frac{\epsilon}{K^2 r_G}}\right],
\end{aligned}
\tag{20.81}
$$

or

$$\tau \approx \frac{\epsilon}{Kc} = \frac{r - r_G}{Kc} \qquad \epsilon \ll r_G. \tag{20.82}$$

From the result (20.82) we conclude that the observer reaches the point $r = r_G$ after the lapse of a finite proper time according to his own clock.

Thus the singularity at $r = r_G$ is not a real unphysical singularity, but merely a coordinate singularity that is a consequence of the choice of the coordinate system. Let us now assume that, during the radial fall, the observer is sending light signals to a distant counterpart at precisely regular time intervals according to his own clock. Using the definition of the differential of the proper time (17.5), the differential of the coordinate time is given by

$$dt = \frac{d\tau}{\sqrt{g_{00}}} = d\tau \left(1 - \frac{r_G}{r}\right)^{-1/2}, \tag{20.83}$$

and from the point of view of the distant receiver the light signals are red-shifted by a factor proportional to

$$(g_{00})^{-1/2} = \left(1 - \frac{r_G}{r}\right)^{-1/2}. \tag{20.84}$$

Thus, although the traveling observer passes the point $r = r_G$ after the lapse of a finite proper time, from the point of view of the distant receiver the time intervals between light signals become longer and longer as the traveling observer approaches the point $r = r_G$. The distant receiver never sees the traveling observer after his passing below the point $r = r_G$, which is in line with the initial assertion that neither matter nor energy (including light signals) can escape from the region with $r < r_G$.

The lack of communication with the outside world is a basic property of black holes. The boundary $r = r_G$ is called an *event horizon*. Just as the curved Earth creates a horizon limiting the range of vision of an ocean navigator, the strongly curved geometry in the vicinity of a black hole creates an event horizon hiding the space-time of the interior of the region with $r \leq r_G$ from the external observer. Nevertheless, the black holes do exert influence on the external observers by their gravitational effects on external bodies arising from the Schwarzschild metric for $r > r_G$.

Another interesting feature of the black holes is that inside the region $r < r_G$ there can be no static bodies. Only the dynamic bodies can exist inside the region $r < r_G$. This property is obvious from the unphysical negative result for the square of the world line element of a body at rest, for which we have $dr = d\theta = d\varphi = 0$, i.e.,

$$ds^2 = c^2 d\tau^2 = c^2 \left(1 - \frac{r_G}{r}\right) dt^2 < 0. \tag{20.85}$$

At this point we want to examine how bodies move inside a black hole. In particular it is of interest to find out if they are moving toward or away from the central mass. We first note that by virtue of the principle of invariance of the speed of light, the world line element ds of a light ray is always equal to zero. The equation $ds = 0$ defines the two *light cones*. As an illustration consider a pseudo-Euclidean metric described by Descartes coordinates, where $dy = dz = 0$. The light cones are then defined by

$$ds^2 = c^2 dt^2 - dx^2 = 0, \qquad (20.86)$$

or

$$x_Q - x_P = \pm c\,(t_Q - t_P). \qquad (20.87)$$

Equation (20.87) is the equation of the cross section of two cones with the y-plane or the z-plane. All world lines of a light ray in four-dimensional space-time, passing through an arbitrary world point P, must lie on these two light cones where $ds = 0$. All other permissible world lines of arbitrary bodies with $ds^2 > 0$ must lie within the two light cones, since otherwise the slope of the actual world line would be larger than c, indicating a body moving faster than the speed of light with $ds^2 < 0$. All world lines within one of the two cones point into the future compared to the world point P, while all world lines in the other two cones point into the past compared to the world point P.

As we have argued before, if we want to explain how bodies are moving inside a black hole, the usual set of four coordinates (t, r, θ, φ) is not adequate and we need to introduce a more suitable new set of coordinates. The simplest way to achieve this goal is to keep the three spatial coordinates and just introduce a new time coordinate w, defined by

$$w = t + \frac{r_G}{c} \ln \left| \frac{r}{r_G} - 1 \right|. \qquad (20.88)$$

From (20.88) we obtain

$$dt = dw - \frac{r_G}{cr} \left(1 - \frac{r_G}{r} \right)^{-1} dr. \qquad (20.89)$$

The square of the original coordinate time differential dt as a function of the new coordinate time differential dw defined by Equation (20.89) is then given by

$$dt^2 = dw^2 - \frac{2r_G}{cr} \left(1 - \frac{r_G}{r} \right)^{-1} dw\,dr$$

$$+ \frac{r_G^2}{c^2 r^2} \left(1 - \frac{r_G}{r} \right)^{-2} dr^2. \qquad (20.90)$$

On the other hand for the light cones of a radial motion, where we have $d\theta = d\varphi = 0$, the result for the metric gives

$$ds^2 = \left(1 - \frac{r_G}{r}\right)c^2 dt^2 - \left(1 - \frac{r_G}{r}\right)^{-1} dr^2 = 0, \qquad (20.91)$$

or

$$dt^2 = \frac{1}{c^2}\left(1 - \frac{r_G}{r}\right)^{-2} dr^2. \qquad (20.92)$$

Substituting (20.92) into (20.90), we can remove the dependency on dt^2 and obtain a quadratic equation in dw/dr as follows:

$$\left(1 - \frac{r_G}{r}\right)\left(\frac{dw}{dr}\right)^2 - \frac{2r_G}{cr}\frac{dw}{dr} - \frac{1}{c^2}\left(1 + \frac{r_G}{r}\right) = 0. \qquad (20.93)$$

The two solutions of the quadratic Equation (20.93) are given by

$$\frac{dw}{dr} = \frac{1}{c}\left(\frac{r_G}{r} \pm 1\right)\left(1 - \frac{r_G}{r}\right)^{-1}, \qquad (20.94)$$

or

$$\frac{dw}{dr} = -\frac{1}{c}, \quad \frac{dw}{dr} = \frac{1}{c}\left(1 + \frac{r_G}{r}\right)\left(1 - \frac{r_G}{r}\right)^{-1}. \qquad (20.95)$$

The first of Equations (20.95) shows that outside the black hole for $r > r_G$, some world lines have a decreasing coordinate distance r with increasing coordinate time w, i.e., the light rays move toward the central mass. The second of Equations (20.95) for $r > r_G$ shows that some world lines have increasing r with increasing coordinate time w, i.e., the light rays move away from the central mass. On the other hand, inside the black hole for $r < r_G$ both Equations (20.95) show that all world lines have decreasing coordinate distance r with increasing coordinate time w, i.e., the light rays always move toward the central mass. Based on the earlier conclusion that all other permissible world lines of arbitrary bodies with $ds^2 > 0$ must lie within the two light cones, we see that all matter and energy (including the light signals) within a black hole for $r < r_G$ can only move toward the central mass and can never escape from the region where $r < r_G$.

The preceding discussion of the properties of black holes would be only academic unless there were reasons to believe that such objects exist in the universe. It is generally believed today that black holes do exist and that they are created by gravitational collapse in the final stage of the evolution of massive stars with a mass greater than 10 times the solar mass. This type of black hole is expected to have a mass in the range of two to three solar masses. Some stellar objects that may be candidates for this type of black hole have already been studied. Super-massive black holes comprising thousands, millions, or billions of solar masses may also exist. It has also been suggested that in the early stages of the creation of the universe, because of the high density of the hot matter, some small material objects could have been squeezed sufficiently to form the so-called mini black holes.

Part V

Elements of Cosmology

► Chapter 21

The Robertson–Walker Metric

In the previous two chapters we have shown that the general theory of relativity provides the solutions for the space-time structure created by any given matter distribution. Thus, if we can specify the average distribution of matter in the entire universe, the general theory of relativity provides the solution for the average space-time structure of the entire universe. The study of such a solution for the average space-time structure of the entire universe is a part of the subject of *cosmology*.

21.1 | Introduction and Basic Observations

In the present chapter we study phenomena on a cosmological scale. For that purpose we need to develop a suitable model of the universe and to make suitable assumptions about the physical processes that are dominant on the cosmological scale. The two basic observations that allow us to address the large-scale structure of the universe are the expansion of the universe (the Hubble law) and the cosmic microwave background radiation.

The first basic observation that allows us to address the large-scale structure of the universe is the discovery that the spectral lines of distant galaxies are shifted toward the red end of the spectrum. If we interpret this red shift as a Doppler shift, this observation leads to the conclusion that all the distant galaxies are receding from us. Thus we conclude that the universe as a whole is expanding. This expansion implies a finite age of the universe. In other words, by observing the sky we can only see stars

that are close enough for the radiation originating from them to reach us in
such a finite time.

It has been shown by Hubble that the velocity of the distant galaxies v
increases linearly with their distance r. This proportionality is known as
the *Hubble law*. Using simple Euclidean geometry, the Hubble law can be
formulated as follows:

$$\vec{v} = H\vec{r} \quad \left(H = (55 \pm 7)\frac{\text{km}}{\text{s}}\frac{1}{\text{Mpc}}\right), \tag{21.1}$$

where the quantity H is called the *Hubble constant*. The Hubble constant is
a constant in a sense that it does not depend on the magnitude or the direction
of the vector $\vec{r}$. However, it may depend on time, and the numerical value
given in (21.1) is its present-time value. According to the Hubble law the
universe is in uniform expansion, which means that there are no privileged
positions in the universe and that an observer traveling with any galaxy sees
the surrounding galaxies as receding from him. If we observe from Earth
two different distant galaxies G_1 and G_2, then according to the Hubble law
their respective velocities observed from our galaxy are given by

$$\vec{v}_1 = H\vec{r}_1, \quad \vec{v}_2 = H\vec{r}_2. \tag{21.2}$$

The relative velocity of the galaxy G_1 as observed from the galaxy G_2,
denoted by V, is then obtained as follows:

$$\vec{V} = \vec{v}_1 - \vec{v}_2 = H(\vec{r}_1 - \vec{r}_2) = H\vec{R}, \tag{21.3}$$

where $\vec{R} = \vec{r}_1 - \vec{r}_2$ is the relative distance between the two galaxies G_1
and G_2. Thus an observer traveling with galaxy G_2 sees galaxy G_1, and
indeed any other galaxy, as receding from him. Although we have used
nonrelativistic approximation in Euclidean geometry and ignored the time
dependence of the Hubble constant, these general conclusions are nonethe-
less valid, i.e., the observer traveling with each galaxy sees all the other
distant galaxies as receding from him.

Since the distant galaxies are receding from us with a velocity directly
proportional to their distance, there must be a point at which each of them
will approach the speed of light and the relativistic effects will become dominant. The set of all those points is called the *world horizon*. The radius of the
world horizon is approximately equal to $r_{WH} = c/H = 2 \times 10^{10}$ lightyears,
if the present-day value of the Hubble constant is used. The galaxies that
are located at or beyond the world horizon are invisible to us.

The second basic observation that allows us to address the large-scale
structure of the universe is the discovery of the extremely isotropic cosmic
microwave background radiation. This radiation has the same intensity in
all directions with a precision better than one part in a thousand. Regardless
of the origin of the radiation, this observation is convincing evidence for the

assumption that any acceptable model of the universe has to be an isotropic model.

Another interesting feature of the microwave background radiation is its relatively high energy density. It is the component of diffuse radiation in the universe with by far the highest energy density. The universe today is matter dominated, but there are reasons to believe that the universe was radiation dominated in the early phases of its history and that the observed microwave background radiation originates from these early phases of the history of the universe.

21.2   Metric Definition and Properties

The basic observations described in the previous section indicate that the large-scale structure of the universe is both homogeneous and isotropic. This assumption is often called the *cosmological principle*. We imagine that the matter in the universe is on the large-scale evenly distributed in the form of a *cosmic fluid*, with no shear-viscous, bulk-viscous, or heat-conductive features. This is a good approximation of the actual universe as long as we take the large-scale point of view. Clearly, on a smaller scale, the matter in the universe is unevenly distributed and the universe is highly nonhomogeneous.

We denote an observer at rest with respect to the cosmic fluid as a *fundamental observer*. As the universe expands, the cosmic fluid takes part in the expansion and the fundamental observer is co-moving with the fluid. Every co-moving observer in the cosmic fluid sees the same isotropic and homogeneous image of the universe. The objective of the present section is to define the space-time metric in the co-moving frame of reference. We cannot use the static Schwarzschild metric to describe the expanding universe, which is certainly not static. We need a nonstatic, homogeneous, and isotropic metric to describe the entire universe.

Let us take the proper time measured by the fundamental observer co-moving with the cosmic fluid as the time coordinate, which we here denote by w. In such a case the space-time metric has the following general form:

$$ds^2 = c^2 dw^2 + 2cg_{0\alpha} dw dx^\alpha + g_{\alpha\beta} dx^\alpha dx^\beta. \qquad (21.4)$$

Applying the condition of isotropy and spherical symmetry to the metric (21.4), we obtain

$$ds^2 = c^2 dw^2 - D(r,w) dr^2 - cE(r,w) dr dw$$
$$- F(r,w)(d\theta^2 + \sin^2\theta d\varphi^2). \qquad (21.5)$$

In Equation (21.5) it should be noted that we have chosen the **dimensionless** radial coordinate $r = \mathcal{R}/L$, where $\mathcal{R}$ is the usual radial coordinate with the dimension of length and L is a yet-unspecified reference length of the universe. The functions $D(r, w)$ and $F(r, w)$ therefore have the dimension of the square of length and the function $E(r, w)$ has the dimension of length to secure the proper dimension of the metric (21.5). Let us now use the geodesic equations of a particle given by (11.32), in the form

$$\frac{du_j}{ds} = \frac{1}{2}\frac{\partial g_{lk}}{\partial x^j}u^l u^k. \tag{21.6}$$

For a particle at rest in the co-moving frame of reference we have

$$u^0 = 1, \quad u^\alpha = 0 \quad (\alpha = 1, 2, 3). \tag{21.7}$$

Substituting (21.7) into (21.6) and using $g_{00} = c^2$, we obtain

$$\frac{du_j}{ds} = \frac{d}{ds}(g_{jl}u^l) = \frac{d}{ds}\left(g_{j0}u^0\right)$$

$$= \frac{dw}{ds}\frac{dg_{j0}}{dw} = \frac{1}{2}\frac{\partial g_{00}}{\partial x^j} = 0. \tag{21.8}$$

Thus we obtain

$$\frac{dg_{j0}}{dw} = 0 \Rightarrow \frac{d}{dw}E(r, w) = 0. \tag{21.9}$$

From (21.9) we conclude that $E = E(r)$ is not a function of time coordinate w, and the metric (21.5) becomes

$$ds^2 = c^2 dw^2 - D(r, w)dr^2 - cE(r)drdw$$
$$- F(r, w)(d\theta^2 + \sin^2\theta d\varphi^2). \tag{21.10}$$

The term proportional to $drdw$ can be eliminated by introducing a new time coordinate:

$$t = w - \frac{1}{2c}\int^r E(r)dr \Rightarrow dw = dt + \frac{1}{2c}E(r)dr. \tag{21.11}$$

Using (21.11), we may write

$$c^2 dw^2 = c^2 dt^2 + cE(r)drdt + \frac{1}{4}[E(r)]^2 dr^2, \tag{21.12}$$

and

$$-cE(r)drdw = -cE(r)drdt - \frac{1}{2}[E(r)]^2 dr^2. \tag{21.13}$$

Substituting (21.12) and (21.13) into (21.10), we obtain

$$ds^2 = c^2 dt^2 - A(r,t)dr^2 - B(r,t)\left(d\theta^2 + \sin^2\theta d\varphi^2\right) \qquad (21.14)$$

where

$$g_{11} = A(r,t) = D[r,w(r,t)] + \frac{1}{2}[E(r)]^2$$

$$g_{22} = B(r,t) = F[r,w(r,t)]. \qquad (21.15)$$

The functions $A(r,t)$ and $B(r,t)$ defined by (21.15) have the dimension of the square of length to secure the proper dimension of the metric (21.14). In order to further specify the functions $A(r,t)$ and $B(r,t)$, we now apply the condition of homogeneity of space-time and consider the following coordinate transformation:

$$z^\alpha = x^\alpha + \varepsilon^\alpha(r,\theta,\varphi) \quad (\alpha = 1,2,3). \qquad (21.16)$$

Using the initial assumption that $\varepsilon^0 \equiv 0$ and Equation (21.16) we obtain

$$\frac{\partial z^\alpha}{\partial x^\beta} = \delta^\alpha_\beta + \frac{\partial \varepsilon^\alpha}{\partial x^\beta}, \quad \frac{\partial z^0}{\partial x^\beta} = 0. \qquad (21.17)$$

The general transformation law for the covariant metric tensor, defined with respect to the systems of coordinates $\{x^k\}$ and $\{z^k\}$ by g_{kn} and $\bar{g}_{jl}$, respectively, is given by

$$g_{kn} = \frac{\partial z^j}{\partial x^k}\frac{\partial z^l}{\partial x^n}\bar{g}_{jl}. \qquad (21.18)$$

Using Equations (21.17) and (21.18) we may write

$$g_{\alpha\beta} = \frac{\partial z^\nu}{\partial x^\alpha}\frac{\partial z^\sigma}{\partial x^\beta}\bar{g}_{\nu\sigma}. \qquad (21.19)$$

Substituting (21.17) into (21.19) we obtain

$$g_{\alpha\beta} = \left(\delta^\nu_\alpha + \frac{\partial \varepsilon^\nu}{\partial x^\alpha}\right)\left(\delta^\sigma_\beta + \frac{\partial \varepsilon^\sigma}{\partial x^\beta}\right)\bar{g}_{\nu\sigma}. \qquad (21.20)$$

Expanding the expression (21.20) and dropping the term quadratic in the derivatives of the small translational parameters ε^α, we obtain

$$g_{\alpha\beta}(x^k) = \bar{g}_{\alpha\beta}(z^k) + \frac{\partial \varepsilon^\nu}{\partial x^\alpha}\bar{g}_{\nu\beta}(z^k) + \frac{\partial \varepsilon^\nu}{\partial x^\beta}\bar{g}_{\nu\alpha}(z^k). \qquad (21.21)$$

The homogeneity of space-time now requires that the metric be invariant with respect to the transformation (21.16), i.e., that we have

$$\bar{g}_{\alpha\beta}(z^k) = g_{\alpha\beta}(z^k) \quad (\alpha,\beta = 1,2,3). \qquad (21.22)$$

Substituting (21.22) into (21.21), we obtain

$$g_{\alpha\beta}(x^k) = \bar{g}_{\alpha\beta}(z^k) + \frac{\partial \varepsilon^\nu}{\partial x^\alpha} g_{\nu\beta}(z^k) + \frac{\partial \varepsilon^\nu}{\partial x^\beta} g_{\nu\alpha}(z^k). \tag{21.23}$$

On the other hand, expanding $g_{\alpha\beta}(z^k)$ into the Taylor series, we may write

$$g_{\alpha\beta}(z^k) = g_{\alpha\beta}(x^k) + \frac{\partial g_{\alpha\beta}}{\partial x^\nu} \varepsilon^\nu. \tag{21.24}$$

Substituting (21.24) into (21.23) and dropping the terms quadratic in the small translational parameters ε^α and their derivatives, we see that only the zeroth-order term of (21.24) contributes to the first-order Equation (21.23). Thus we obtain

$$g_{\alpha\beta}(x^k) = \bar{g}_{\alpha\beta}(z^k) + \frac{\partial \varepsilon^\nu}{\partial x^\alpha} g_{\nu\beta}(x^k) + \frac{\partial \varepsilon^\nu}{\partial x^\beta} g_{\nu\alpha}(x^k), \tag{21.25}$$

or

$$\bar{g}_{\alpha\beta}(z^k) = g_{\alpha\beta}(x^k) - \frac{\partial \varepsilon^\nu}{\partial x^\alpha} g_{\nu\beta}(x^k) - \frac{\partial \varepsilon^\nu}{\partial x^\beta} g_{\nu\alpha}(x^k). \tag{21.26}$$

Using again the homogeneity condition (21.22) and comparing Equations (21.26) and (21.24) with each other, we obtain

$$\frac{\partial g_{\alpha\beta}}{\partial x^\nu} \varepsilon^\nu + \frac{\partial \varepsilon^\nu}{\partial x^\alpha} g_{\nu\beta}(x^k) + \frac{\partial \varepsilon^\nu}{\partial x^\beta} g_{\nu\alpha}(x^k) = 0. \tag{21.27}$$

For $\alpha = \beta = 1$ and using the metric (21.14) with (21.15), Equation (21.27) becomes

$$\frac{\partial A}{\partial r} \varepsilon^1 + 2 \frac{\partial \varepsilon^1}{\partial r} A = 0. \tag{21.28}$$

For $\alpha = 1$ and $\beta = 2$ and using metric (21.14) with (21.15), Equation (21.27) becomes

$$\frac{\partial \varepsilon^2}{\partial r} B + \frac{\partial \varepsilon^1}{\partial \theta} A = 0. \tag{21.29}$$

Finally for $\alpha = \beta = 2$ and using the metric (21.14) with (21.15), Equation (21.27) becomes

$$\frac{\partial B}{\partial r} \varepsilon^1 + 2 \frac{\partial \varepsilon^2}{\partial \theta} B = 0. \tag{21.30}$$

From the differential Equations (21.29) and (21.30) it is possible to eliminate ε^2. We first rewrite (21.29) and (21.30) as

$$-\frac{\partial \varepsilon^2}{\partial r} = \frac{\partial \varepsilon^1}{\partial \theta} \frac{A}{B}$$

$$\frac{\partial \varepsilon^2}{\partial \theta} = -\frac{1}{2B} \frac{\partial B}{\partial r} \varepsilon^1 = \frac{\partial}{\partial r}\left(\ln B^{-1/2}\right)\varepsilon^1. \tag{21.31}$$

From (21.31) we may write

$$-\frac{\partial^2 \varepsilon^2}{\partial r \partial \theta} = \frac{\partial^2 \varepsilon^1}{\partial \theta^2} \frac{A}{B}$$

$$\frac{\partial^2 \varepsilon^2}{\partial r \partial \theta} = \frac{\partial^2}{\partial r^2}\left(\ln B^{-1/2}\right)\varepsilon^1 + \frac{\partial}{\partial r}\left(\ln B^{-1/2}\right)\frac{\partial \varepsilon^1}{\partial r}. \tag{21.32}$$

Addition of the two Equations (21.32) gives

$$\frac{\partial^2 \varepsilon^1}{\partial \theta^2}\frac{A}{B} + \frac{\partial^2}{\partial r^2}\left(\ln B^{-1/2}\right)\varepsilon^1 + \frac{\partial}{\partial r}\left(\ln B^{-1/2}\right)\frac{\partial \varepsilon^1}{\partial r} = 0. \tag{21.33}$$

Now, using Equations (21.28) and (21.30), we may also write

$$\frac{1}{A}\frac{\partial A}{\partial r} = \frac{\partial}{\partial r}(\ln A) = -\frac{2}{\varepsilon^1}\frac{\partial \varepsilon^1}{\partial r}$$

$$\frac{1}{B}\frac{\partial B}{\partial r} = \frac{\partial}{\partial r}(\ln B) = -\frac{2}{\varepsilon^1}\frac{\partial \varepsilon^2}{\partial \theta}. \tag{21.34}$$

As the small translation parameters ε^α are by definition independent of the time coordinate t, we conclude that $\partial_r(\ln A)$ and $\partial_r(\ln B)$ are not the functions of the time coordinate t, either. But the functions $A(r,t)$ and $B(r,t)$ are dependent on the time coordinate t. This is only possible if the functions $A(r,t)$ and $B(r,t)$ are factorized as follows:

$$A(r,t) = a(r)R^2(t), \quad B(r,t) = b(r)R^2(t), \tag{21.35}$$

where $R(t)$ is some yet unspecified function of the time coordinate t that has the dimension of length. This function in some sense plays the role of the radius of the universe. The function $R(t)$ should not be confused with the Ricci scalar R used in the gravitational field equations, and whenever there is a risk of confusion the distinction between the two will be explicitly stated. On the other hand, we note that the dimensionless radial coordinate r is not uniquely defined. The metric (21.14) is invariant with respect to the transformation from the radial coordinate r to some new radial coordinate ρ, defined by

$$r = \int^{\rho} F(\rho)d\rho \Rightarrow dr = F(\rho)d\rho, \tag{21.36}$$

where $F(\rho)$ is some arbitrary function of the new radial coordinate ρ. Indeed, if we substitute (21.36) into the metric (21.14), we obtain

$$ds^2 = c^2dt^2 - \bar{A}(\rho,t)d\rho^2 - \bar{B}(\rho,t)\left(d\theta^2 + \sin^2\theta d\varphi^2\right) \tag{21.37}$$

where

$$g_{11} = \bar{A}(\rho,t) = A\left[\int^{\rho} F(\rho)d\rho, t\right][F(\rho)]^2$$

$$g_{22} = \bar{B}(\rho,t) = B\left[\int^{\rho} F(\rho)d\rho, t\right]. \tag{21.38}$$

The metric (21.37) has the same form as the metric (21.14) and they are fully equivalent to each other. Thus we are free to make an arbitrary choice of the form of one of the functions $a(r)$ or $b(r)$. If we set $b(r) = r^2$, the metric (21.14) becomes

$$ds^2 = c^2dt^2 - R^2(t)a(r)dr^2 - R^2(t)r^2\left(d\theta^2 + \sin^2\theta d\varphi^2\right). \tag{21.39}$$

From the metric (21.39) we see that the surface area of a sphere with the relative radius r is equal to $4\pi r^2 R^2(t)$. If, at a conveniently chosen time instant t, we have $R(t) = L$, then the surface area of a sphere with the relative radius $r = R/L$ is equal to $4\pi R^2$ and the angular coordinates θ and φ become the usual spherical angular coordinates. In order to specify the function $a = a(r)$, we use Equation (21.28) and the factorized form of the function $A(r,t)$ given by (21.35) to obtain

$$\frac{\partial}{\partial r}\left(\ln\varepsilon^1\right) = \frac{\partial}{\partial r}\left(\ln A^{-1/2}\right) = \frac{\partial}{\partial r}\left(\ln a^{-1/2}\right). \tag{21.40}$$

The solution of the differential Equation (21.40) can be written as

$$\varepsilon^1(r,\theta,\varphi) = e(\theta,\varphi)a^{-1/2}(r). \tag{21.41}$$

Substituting the factorized form of the functions $A(r,t)$ and $B(r,t)$, given by (21.35) with $b(r) = r^2$, into Equation (21.33), we obtain

$$\frac{\partial^2\varepsilon^1}{\partial\theta^2}\frac{a}{r^2} = \frac{\partial^2}{\partial r^2}(\ln r)\,\varepsilon^1 + \frac{\partial}{\partial r}(\ln r)\frac{\partial\varepsilon^1}{\partial r}, \tag{21.42}$$

or

$$\frac{\partial^2 \varepsilon^1}{\partial \theta^2} = \frac{r^2}{a} \left(\frac{1}{r} \frac{\partial \varepsilon^1}{\partial r} - \frac{1}{r^2} \varepsilon^1 \right) = \frac{r^2}{a} \frac{\partial}{\partial r} \left(\frac{\varepsilon^1}{r} \right). \tag{21.43}$$

Substituting the solution (21.41) into (21.43) we further obtain

$$\frac{1}{e} \frac{\partial^2 e}{\partial \theta^2} = \frac{r^2}{a^{1/2}} \frac{\partial}{\partial r} \left(\frac{1}{r a^{1/2}} \right) = C. \tag{21.44}$$

The expression on the right-hand side of Equation (21.44) is a function of the angular variables θ and φ only, while the expression on the left-hand side is a function of the radial variable r only. Thus both sides of Equation (21.44) are equal to a constant that we denote by C. In order to determine the value of C, we use (21.44) to write

$$\frac{\partial^2 e}{\partial \theta^2} = C e. \tag{21.45}$$

If the translation (21.16) is chosen to be along the polar axis (z-axis) in the corresponding Descartes coordinates, we have $e(\theta, \varphi) \sim \cos\theta$. This determines the value of the constant C to be $C = -1$. Thus we obtain the differential equation for the function $a = a(r)$ in the form

$$\frac{r^2}{a^{1/2}} \frac{\partial}{\partial r} \left(\frac{1}{r a^{1/2}} \right) = -1. \tag{21.46}$$

The solution of the differential Equation (21.46) is

$$a(r) = \frac{1}{1 - kr^2}. \tag{21.47}$$

It is easy to verify by direct calculation that (21.47) satisfies the differential Equation (21.46):

$$r^2 \sqrt{1 - k r^2} \frac{\partial}{\partial r} \left(\frac{\sqrt{1 - kr^2}}{r} \right) = -1. \tag{21.48}$$

Substituting (21.47) into (21.39) we obtain the final result for the *Robertson–Walker Metric* in the form

$$ds^2 = c^2 dt^2 - R^2(t) \left(\frac{dr^2}{1 - k r^2} + r^2 d\theta^2 + r^2 \sin^2\theta d\varphi^2 \right) \tag{21.49}$$

where $R(t)$ is a dynamic function of the time coordinate t with the dimension of length, which will be calculated as a solution to the gravitational field equations, and k is the so-called *curvature constant*, which describes the geometry of the three-dimensional space at any particular time instant. The positive values of the curvature constant ($k > 0$) correspond to the

three-dimensional space with a positive curvature, the zero value ($k = 0$)
corresponds to the flat three-dimensional space, and the negative values ($k < 0$) correspond to the three-dimensional space with a negative
curvature. The value of the curvature constant k can always be taken as
$+1, 0$, or -1 by a suitable rescaling of the radial coordinate r.

The infinitesimal element of the proper distance between two arbitrary
galaxies given by cdt can be calculated from the condition $ds = 0$ using
the Robertson–Walker metric (21.49) as follows:

$$cdt = R(t)d\sigma, \quad d\sigma^2 = \frac{dr^2}{1 - kr^2} + r^2 d\theta^2 + r^2 \sin^2 \theta d\varphi^2 \qquad (21.50)$$

where $d\sigma^2$ is the metric of the three-dimensional space for a fixed value of
the time coordinate t that is usually called the *cosmic time*. The cosmic time
is a universal time equal for all observers at rest with respect to the local
matter. Using Equation (21.50) we may write $c\Delta t = R(t)\Delta\sigma$ for finite
proper distances. All measurements are made at the same epoch t. The
radial coordinate r is a co-moving coordinate and it remains fixed for each
galaxy. The angular coordinates θ and φ also remain fixed for the isotropic
motion of each galaxy. Thus the spatial metric $d\sigma^2$ remains fixed and the
proper distance between two galaxies is only scaled by the function $R(t)$
as the cosmic time t varies. The function $R(t)$ with the dimension of length
is therefore called the *scale radius*, and it increases or decreases as the
universe expands or contracts, respectively.

21.3 The Hubble Law

In the present section we discuss the Hubble law in the framework of
the Robertson–Walker geometry. It has been concluded before that each
galaxy has similar coordinates (r, θ, φ). Let us now assume that our
own galaxy, being an approximately co-moving object, lies at the spatial
origin $r = 0$ and that some other distant galaxy lies at some radial coordinate distance r. Because of the homogeneity of space-time this choice of
coordinates does not place our own galaxy in the center of the universe,
since any galaxy can be chosen to have $r = 0$. This particular choice is only
made for convenience. The proper radial distance D_r between our galaxy
at $r = 0$ and the distant galaxy, at some radial coordinate distance r and
at a given cosmic time t, can be calculated using the Robertson–Walker
metric as

$$D_r = R(t) \int_0^r \frac{dr}{\sqrt{1 - kr^2}}. \qquad (21.51)$$

The integral (21.51) is elementary and gives

$$D_r = \begin{cases} R(t)\arcsin r & (k = 1) \\ R(t)\, r & (k = 0) \\ R(t)\operatorname{arcsinh} r & (k = -1) \end{cases} \qquad (21.52)$$

where it should be noted that $r = \mathcal{R}/L$ is the dimensionless radial coordinate. Thus the proper distance is proportional to the scale radius $R(t)$, which changes with time. Keeping in mind that the radial coordinate r is a fixed co-moving coordinate, the proper velocity is obtained by differentiating the proper distance D_r with respect to the cosmic time t, i.e.,

$$v_r = \dot{D}_r = \dot{R}(t) \int_0^r \frac{dr}{\sqrt{1 - kr^2}} = \frac{\dot{R}}{R} D_r, \qquad (21.53)$$

where the dots denote the time differentiation. Thus we obtain the Hubble law saying that at any given cosmic time t the speed of any distant galaxy relative to our own galaxy is proportional to its proper distance from our galaxy. The Hubble constant is given by

$$H(t) = \frac{\dot{R}(t)}{R(t)}, \qquad (21.54)$$

and we see that the Hubble constant is indeed a function of the cosmic time as we have anticipated earlier in this chapter. The quantity H_0 measured by the astronomers is the value of the Hubble constant at the present epoch, i.e., for $t = t_0$. It should be noted that the proper distance is not a directly measurable quantity, and it can only be measured indirectly by measurements of some other quantities such as red shifts.

21.4 The Cosmological Red Shifts

In the present section we discuss cosmological red shifts as one of the means of measuring the proper distances to distant galaxies and test the validity of the Hubble law in the framework of the Robertson–Walker geometry. Let us consider a distant galaxy at some relative radial coordinate distance $r = r_d$ emitting two light wave crests at cosmic times t_d and $t_d + \Delta t_d$ toward our own galaxy situated at $r = 0$. The two wave crests are received in our galaxy at cosmic times t_0 and $t_0 + \Delta t_0$. For the radial motion of light the Robertson–Walker metric gives

$$ds^2 = c^2 dt^2 - R^2(t)\frac{dr^2}{1 - kr^2} = 0. \qquad (21.55)$$

Thus we may write

$$\frac{dt}{R(t)} = \pm \frac{1}{c} \frac{dr}{1 - kr^2}. \tag{21.56}$$

Since the beam of light is moving toward us, the radial coordinate r decreases as the time coordinate t increases along the null geodesic, and it is appropriate to use a minus sign in Equation (21.56). Thus we may integrate Equation (21.56) to obtain

$$\int_{t_d}^{t_0} \frac{dt}{R(t)} = \frac{1}{c} \int_0^{r_d} \frac{dr}{1 - kr^2} \tag{21.57}$$

and

$$\int_{t_d + \Delta t_d}^{t_0 + \Delta t_0} \frac{dt}{R(t)} = \frac{1}{c} \int_0^{r_d} \frac{dr}{1 - kr^2}. \tag{21.58}$$

For all types of radiation received from distant galaxies, the time intervals Δt_d and Δt_0 are tiny fractions of a second, and over that time R(t) remains effectively constant. Subtracting Equation (21.57) from Equation (21.58), we obtain

$$\frac{\Delta t_0}{R(t_0)} - \frac{\Delta t_d}{R(t_d)} = 0 \Rightarrow \frac{\Delta t_0}{\Delta t_d} = \frac{R(t_0)}{R(t_d)}. \tag{21.59}$$

The observed wavelength λ_0 and the emitted wavelength λ_d are related to the time intervals Δt_0 and Δt_d by the following definitions:

$$\lambda_0 = c \Delta t_0, \quad \lambda_d = c \Delta t_d. \tag{21.60}$$

Thus the red shift of the received light waves can be obtained in terms of the function $R(t)$ as follows:

$$z = \frac{\lambda_0 - \lambda_d}{\lambda_d} = \frac{R(t_0)}{R(t_d)} - 1. \tag{21.61}$$

In the expanding universe we have $R(t_0) > R(t_d)$ and the red shift z is positive in agreement with the empirical observations. The red shift (21.61) is a consequence of the light traveling in curved space-time and is not a result of the Doppler effect. This red shift is called the *cosmological red shift*. Most observed cosmological red shifts are rather small and t_d is relatively close to t_0. It is therefore possible to expand $R(t_d)$ in a Taylor series around t_0 as follows:

$$R(t_d) = R(t_0) + (t_d - t_0)\dot{R}(t_0) + \frac{1}{2}(t_d - t_0)^2 \ddot{R}(t_0) + \cdots \tag{21.62}$$

or

$$R(t_d) = R(t_0) \left[1 + H_0(t_d - t_0)\dot{R}(t_0) - \frac{1}{2}q_0 H_0^2(t_d - t_0)^2 + \cdots \right]$$

$$(21.63)$$

where H_0 is the present value of the Hubble constant given by

$$H_0 = \frac{\dot{R}(t_0)}{R(t_0)},$$

$$(21.64)$$

and q_0 is a dimensionless *deceleration parameter* given by

$$q_0 = -\frac{\ddot{R}(t_0)R(t_0)}{\dot{R}^2(t_0)}.$$

$$(21.65)$$

The deceleration parameter q_0 is positive when $\ddot{R}(t)$ is negative, i.e., when the expansion of the universe is slowing down. Substituting (21.63) into (21.61), we obtain

$$z \approx \left[1 + H_0(t_d - t_0)\dot{R}(t_0) - \frac{1}{2}q_0 H_0^2(t_d - t_0)^2 \right]^{-1} - 1 \qquad (21.66)$$

or

$$z \approx H_0(t_0 - t_d) + \left(1 + \frac{1}{2}q_0 \right) H_0^2(t_0 - t_d)^2.$$

$$(21.67)$$

This formula is sometimes very useful, but we must keep in mind that it is only valid for small cosmological red shifts where t_d is relatively close to t_0.

When we observe a galaxy with a red shift $z = 1$, it means that the scale radius $R(t_d)$ of the universe at the cosmic time t_d when the radiation was emitted was one-half of the present scale radius $R(t_0)$. In other words, the size of the universe at the time t_d was half of the present size of the universe. Unfortunately, we do not know the cosmic time t_d when the radiation was emitted. If we did, we could directly measure the function $R(t)$. We therefore need some theory of the cosmic dynamics in order to determine the scale radius $R(t)$. Such a cosmic dynamics theory is the subject of the next chapter.

Chapter 22

Cosmic Dynamics

The nonstatic models of the universe based on the Robertson–Walker metric (21.49) are described by their scale radius $R(t)$ and the curvature constant k. The analysis in the previous chapter did not determine the scale radius $R(t)$ as a function of the cosmic time t. The knowledge of the function $R(t)$ is essential for determining the rate of expansion of the universe and other physical properties of the expanding universe. In order to find the solution for $R(t)$, we need a theory of cosmic dynamics based on the gravitational field equations. We therefore combine the homogeneous isotropic Robertson–Walker metric with the gravitational field equations to obtain the dynamic equations satisfied by the scale radius $R(t)$. These equations are called the *Friedmann equations*.

22.1 The Einstein Tensor

In the present section we use the Robertson–Walker metric (21.49) given by

$$ds^2 = c^2 dt^2 - R^2(t) \left(\frac{dr^2}{1 - kr^2} + r^2 d\theta^2 + r^2 \sin^2 \theta d\varphi^2 \right). \qquad (22.1)$$

The covariant metric tensor for the Robertson–Walker metric is given by the following matrix:

$$[g_{mn}] = \begin{bmatrix} 1 & 0 & 0 & 0 \\ 0 & -R^2(1 - kr^2)^{-1} & 0 & 0 \\ 0 & 0 & -R^2 r^2 & 0 \\ 0 & 0 & 0 & -R^2 r^2 \sin^2 \theta \end{bmatrix}. \qquad (22.2)$$

The contravariant metric tensor for the Robertson–Walker metric is then given by

$$[g_{mn}] = \begin{bmatrix} 1 & 0 & 0 & 0 \\ 0 & -R^{-2}(1-kr^2) & 0 & 0 \\ 0 & 0 & -R^{-2}r^{-2} & 0 \\ 0 & 0 & 0 & -R^{-2}r^{-2}\sin^{-2}\theta \end{bmatrix}. \quad (22.3)$$

Using the matrix (22.2) we may write

$$g_{00} = 1, \quad g_{11} = -\frac{R^2}{1-kr^2},$$

$$g_{22} = -R^2 r^2, \quad g_{33} = -R^2 r^2 \sin^2\theta. \quad (22.4)$$

The coordinate differentials of the metric tensor components (22.4) can be calculated as

$$\partial_k g_{00} = 0 \quad (k = 0, 1, 2, 3),$$

$$\partial_0 g_{11} = -\frac{2R\dot{R}}{1-kr^2}, \quad \partial_r g_{11} = -\frac{2krR^2}{(1-kr^2)^2},$$

$$\partial_\theta g_{11} = \partial_\varphi g_{11} = 0,$$

$$\partial_0 g_{22} = -2R\dot{R}r^2, \quad \partial_r g_{22} = -2R^2 r,$$

$$\partial_\theta g_{22} = \partial_\varphi g_{22} = 0,$$

$$\partial_0 g_{33} = -2R\dot{R}r^2\sin^2\theta, \quad \partial_r g_{33} = -2R^2 r\sin^2\theta,$$

$$\partial_\theta g_{33} = -2R^2 r^2\sin\theta\cos\theta, \quad \partial_\varphi g_{33} = 0. \quad (22.5)$$

In the results (22.5) and in the following calculations, we temporarily define by dots the derivatives of the scale radius $R(t)$ with respect to the temporal coordinate $x^0 = ct$ rather than with respect to cosmic time t to simplify the calculations and to make the results compatible with the results obtained elsewhere in the literature using the units with $c = 1$. Thus we have

$$\dot{R}(t) = \frac{dR(t)}{dx^0} = \frac{1}{c}\frac{dR(t)}{dt}. \quad (22.6)$$

The Christoffel symbols of the first kind for the metric (22.1) can now be calculated using the definition (9.28), i.e.,

$$\Gamma_{j,kn} = \frac{1}{2}\left(\frac{\partial g_{jk}}{\partial x^n} + \frac{\partial g_{nj}}{\partial x^k} - \frac{\partial g_{kn}}{\partial x^j}\right). \quad (22.7)$$

The results for the Christoffel symbols of the first kind for the metric (22.1) are summarized in the following list:

$$\Gamma_{0,00} = \frac{1}{2}(\partial_0 g_{00} + \partial_0 g_{00} - \partial_0 g_{00}) = 0$$

$$\Gamma_{0,01} = \Gamma_{0,10} = \frac{1}{2}(\partial_r g_{00} + \partial_0 g_{10} - \partial_0 g_{01}) = 0$$

$$\Gamma_{0,02} = \Gamma_{0,20} = \frac{1}{2}(\partial_\theta g_{00} + \partial_0 g_{20} - \partial_0 g_{02}) = 0$$

$$\Gamma_{0,03} = \Gamma_{0,30} = \frac{1}{2}(\partial_\varphi g_{00} + \partial_0 g_{30} - \partial_0 g_{03}) = 0$$

$$\Gamma_{0,11} = \frac{1}{2}(\partial_r g_{01} + \partial_r g_{10} - \partial_0 g_{11}) = +\frac{R\dot{R}}{1 - kr^2}$$

$$\Gamma_{0,12} = \Gamma_{0,21} = \frac{1}{2}(\partial_\theta g_{01} + \partial_r g_{20} - \partial_0 g_{12}) = 0$$

$$\Gamma_{0,13} = \Gamma_{0,31} = \frac{1}{2}(\partial_\varphi g_{01} + \partial_r g_{30} - \partial_0 g_{13}) = 0$$

$$\Gamma_{0,22} = \frac{1}{2}(\partial_\theta g_{02} + \partial_\theta g_{20} - \partial_0 g_{22}) = +R\dot{R}\,r^2$$

$$\Gamma_{0,23} = \Gamma_{0,32} = \frac{1}{2}(\partial_\varphi g_{02} + \partial_\theta g_{30} - \partial_0 g_{23}) = 0$$

$$\Gamma_{0,33} = \frac{1}{2}(\partial_\varphi g_{03} + \partial_\varphi g_{30} - \partial_0 g_{33}) = +R\dot{R}\,r^2 \sin^2\theta$$

$$\Gamma_{1,00} = \frac{1}{2}(\partial_0 g_{10} + \partial_0 g_{01} - \partial_r g_{00}) = 0$$

$$\Gamma_{1,01} = \Gamma_{1,10} = \frac{1}{2}(\partial_r g_{10} + \partial_0 g_{11} - \partial_r g_{01}) = -\frac{R\dot{R}}{1 - kr^2}$$

$$\Gamma_{1,02} = \Gamma_{1,20} = \frac{1}{2}(\partial_\theta g_{10} + \partial_0 g_{21} - \partial_r g_{02}) = 0$$

$$\Gamma_{1,03} = \Gamma_{1,30} = \frac{1}{2}(\partial_\varphi g_{10} + \partial_0 g_{31} - \partial_r g_{03}) = 0$$

$$\Gamma_{1,11} = \frac{1}{2}(\partial_r g_{11} + \partial_r g_{11} - \partial_r g_{11}) = -\frac{R^2 kr}{(1 - kr^2)^2}$$

$$\Gamma_{1,12} = \Gamma_{1,21} = \frac{1}{2}(\partial_\theta g_{11} + \partial_r g_{21} - \partial_r g_{12}) = 0$$

$$\Gamma_{1,13} = \Gamma_{1,31} = \frac{1}{2}(\partial_\varphi g_{11} + \partial_r g_{31} - \partial_r g_{13}) = 0$$

$$\Gamma_{1,22} = \frac{1}{2}(\partial_\theta g_{12} + \partial_\theta g_{21} - \partial_r g_{22}) = +R^2 r$$

$$\Gamma_{1,23} = \Gamma_{1,32} = \frac{1}{2}(\partial_3 g_{12} + \partial_\theta g_{31} - \partial_r g_{23}) = 0$$

$$\Gamma_{1,33} = \frac{1}{2}(\partial_\varphi g_{13} + \partial_\varphi g_{31} - \partial_r g_{33}) = +R^2 r \sin^2\theta$$

$$\Gamma_{2,00} = \frac{1}{2}(\partial_0 g_{20} + \partial_0 g_{02} - \partial_\theta g_{00}) = 0$$

$$\Gamma_{2,01} = \Gamma_{2,10} = \frac{1}{2}(\partial_r g_{20} + \partial_0 g_{12} - \partial_\theta g_{01}) = 0$$

$$\Gamma_{2,02} = \Gamma_{2,20} = \frac{1}{2}(\partial_\theta g_{20} + \partial_0 g_{22} - \partial_\theta g_{02}) = -R\dot{R}r^2$$

$$\Gamma_{2,03} = \Gamma_{2,30} = \frac{1}{2}(\partial_\varphi g_{20} + \partial_0 g_{32} - \partial_\theta g_{03}) = 0$$

$$\Gamma_{2,11} = \frac{1}{2}(\partial_r g_{21} + \partial_r g_{12} - \partial_\theta g_{11}) = 0$$

$$\Gamma_{2,12} = \Gamma_{2,21} = \frac{1}{2}(\partial_\theta g_{21} + \partial_r g_{22} - \partial_\theta g_{12}) = -R^2 r$$

$$\Gamma_{2,13} = \Gamma_{2,31} = \frac{1}{2}(\partial_\varphi g_{21} + \partial_r g_{32} - \partial_\theta g_{13}) = 0$$

$$\Gamma_{2,22} = \frac{1}{2}(\partial_\theta g_{22} + \partial_\theta g_{22} - \partial_\theta g_{22}) = 0$$

$$\Gamma_{2,23} = \Gamma_{2,32} = \frac{1}{2}(\partial_\varphi g_{22} + \partial_\theta g_{32} - \partial_\theta g_{23}) = 0$$

$$\Gamma_{2,33} = \frac{1}{2}(\partial_\varphi g_{23} + \partial_\varphi g_{32} - \partial_\theta g_{33}) = +R^2 r^2 \sin\theta\cos\theta$$

$$\Gamma_{3,00} = \frac{1}{2}(\partial_0 g_{30} + \partial_0 g_{03} - \partial_\varphi g_{00}) = 0$$

$$\Gamma_{3,01} = \Gamma_{3,10} = \frac{1}{2}(\partial_r g_{30} + \partial_0 g_{13} - \partial_\varphi g_{01}) = 0$$

$$\Gamma_{3,02} = \Gamma_{3,20} = \frac{1}{2}(\partial_\theta g_{30} + \partial_0 g_{23} - \partial_\varphi g_{02}) = 0$$

$$\Gamma_{3,03} = \Gamma_{3,30} = \frac{1}{2}(\partial_\varphi g_{30} + \partial_0 g_{33} - \partial_\varphi g_{03}) = -R\dot{R}r^2 \sin^2\theta$$

$$\Gamma_{3,11} = \frac{1}{2}(\partial_r g_{31} + \partial_r g_{13} - \partial_\varphi g_{11}) = 0$$

$$\Gamma_{3,12} = \Gamma_{3,21} = \frac{1}{2}(\partial_\theta g_{31} + \partial_r g_{23} - \partial_\varphi g_{12}) = 0$$

$$\Gamma_{3,13} = \Gamma_{3,31} = \frac{1}{2}(\partial_\varphi g_{31} + \partial_r g_{33} - \partial_\varphi g_{13}) = -R^2 r \sin^2\theta$$

$$\Gamma_{3,22} = \frac{1}{2}(\partial_\theta g_{32} + \partial_\theta g_{23} - \partial_\varphi g_{22}) = 0$$

$$\Gamma_{3,23} = \Gamma_{3,32} = \frac{1}{2}(\partial_\varphi g_{32} + \partial_\theta g_{33} - \partial_\varphi g_{23}) = -R^2 r^2 \sin\theta \cos\theta$$

$$\Gamma_{3,33} = \frac{1}{2}(\partial_\varphi g_{33} + \partial_\varphi g_{33} - \partial_\varphi g_{33}) = 0. \tag{22.8}$$

The Christoffel symbols of the second kind for the metric (22.1) can be calculated using the definition (9.29), i.e.,

$$\Gamma^p_{kn} = g^{pj}\Gamma_{j,kn}. \tag{22.9}$$

The results for the Christoffel symbols of the second kind for the metric (22.1) are summarized in the following list:

$$\Gamma^0_{00} = g^{0j}\Gamma_{j,00} = g^{00}\Gamma_{0,00} = 0$$

$$\Gamma^0_{01} = \Gamma^0_{10} = g^{0j}\Gamma_{j,01} = g^{00}\Gamma_{0,01} = 0$$

$$\Gamma^0_{02} = \Gamma^0_{20} = g^{0j}\Gamma_{j,02} = g^{00}\Gamma_{0,02} = 0$$

$$\Gamma^0_{03} = \Gamma^0_{30} = g^{0j}\Gamma_{j,03} = g^{00}\Gamma_{0,03} = 0$$

$$\Gamma^0_{11} = g^{0j}\Gamma_{j,11} = g^{00}\Gamma_{0,11} = +\frac{R\dot{R}}{1-kr^2}$$

$$\Gamma^0_{12} = \Gamma^0_{21} = g^{0j}\Gamma_{j,12} = g^{00}\Gamma_{0,12} = 0$$

$$\Gamma^0_{13} = \Gamma^0_{31} = g^{0j}\Gamma_{j,13} = g^{00}\Gamma_{0,13} = 0$$

$$\Gamma^0_{22} = g^{0j}\Gamma_{j,22} = g^{00}\Gamma_{0,22} = +R\dot{R}r^2$$

$$\Gamma^0_{23} = \Gamma^0_{32} = g^{0j}\Gamma_{j,23} = g^{00}\Gamma_{0,23} = 0$$

$$\Gamma^0_{33} = g^{0j}\Gamma_{j,33} = g^{00}\Gamma_{0,33} = +R\dot{R}r^2 \sin^2\theta$$

$$\Gamma^1_{00} = g^{1j}\Gamma_{j,00} = g^{11}\Gamma_{1,00} = 0$$

$$\Gamma^1_{01} = \Gamma^1_{10} = g^{1j}\Gamma_{j,01} = g^{11}\Gamma_{1,01} = +\frac{\dot{R}}{R}$$

$$\Gamma^1_{02} = \Gamma^1_{20} = g^{1j}\Gamma_{j,02} = g^{11}\Gamma_{1,02} = 0$$

$$\Gamma^1_{03} = \Gamma^1_{30} = g^{1j}\Gamma_{j,03} = g^{11}\Gamma_{1,03} = 0$$

$$\Gamma^1_{11} = g^{1j}\Gamma_{j,11} = g^{11}\Gamma_{1,11} = +\frac{kr}{1-kr^2}$$

$$\Gamma^1_{12} = \Gamma^1_{21} = g^{1j}\Gamma_{j,12} = g^{11}\Gamma_{1,12} = 0$$

$$\Gamma^1_{13} = \Gamma^1_{31} = g^{1j}\Gamma_{j,13} = g^{11}\Gamma_{1,13} = 0$$

$$\Gamma^1_{22} = g^{1j}\Gamma_{j,22} = g^{11}\Gamma_{1,22} = -r\left(1 - kr^2\right)$$

$$\Gamma^1_{23} = \Gamma^1_{32} = g^{1j}\Gamma_{j,23} = g^{11}\Gamma_{1,23} = 0$$

$$\Gamma^1_{33} = g^{1j}\Gamma_{j,33} = g^{11}\Gamma_{1,33} = -r\left(1 - kr^2\right)\sin^2\theta$$

$$\Gamma^2_{00} = g^{2j}\Gamma_{j,00} = g^{22}\Gamma_{2,00} = 0$$

$$\Gamma^2_{01} = \Gamma^2_{10} = g^{2j}\Gamma_{j,01} = g^{22}\Gamma_{2,01} = 0$$

$$\Gamma^2_{02} = \Gamma^2_{20} = g^{2j}\Gamma_{j,02} = g^{22}\Gamma_{2,02} = +\frac{\dot{R}}{R}$$

$$\Gamma^2_{03} = \Gamma^2_{30} = g^{2j}\Gamma_{j,03} = g^{22}\Gamma_{2,03} = 0$$

$$\Gamma^2_{11} = g^{2j}\Gamma_{j,11} = g^{22}\Gamma_{2,11} = 0$$

$$\Gamma^2_{12} = \Gamma^2_{21} = g^{2j}\Gamma_{j,12} = g^{22}\Gamma_{2,12} = +\frac{1}{r}$$

$$\Gamma^2_{13} = \Gamma^2_{31} = g^{2j}\Gamma_{j,13} = g^{22}\Gamma_{2,13} = 0$$

$$\Gamma^2_{22} = g^{2j}\Gamma_{j,22} = g^{22}\Gamma_{2,22} = 0$$

$$\Gamma^2_{23} = \Gamma^2_{32} = g^{2j}\Gamma_{j,23} = g^{22}\Gamma_{2,23} = 0$$

$$\Gamma^2_{33} = g^{2j}\Gamma_{j,33} = g^{22}\Gamma_{2,33} = -\sin\theta\cos\theta$$

$$\Gamma^3_{00} = g^{3j}\Gamma_{j,00} = g^{33}\Gamma_{3,00} = 0$$

$$\Gamma^3_{01} = \Gamma^3_{10} = g^{3j}\Gamma_{j,01} = g^{33}\Gamma_{3,01} = 0$$

$$\Gamma^3_{02} = \Gamma^3_{20} = g^{3j}\Gamma_{j,02} = g^{33}\Gamma_{3,02} = 0$$

$$\Gamma^3_{03} = \Gamma^3_{30} = g^{3j}\Gamma_{j,03} = g^{33}\Gamma_{3,03} = +\frac{\dot{R}}{R}$$

$$\Gamma^3_{11} = g^{3j}\Gamma_{j,11} = g^{33}\Gamma_{3,11} = 0$$

$$\Gamma^3_{12} = \Gamma^3_{21} = g^{3j}\Gamma_{j,12} = g^{33}\Gamma_{3,12} = 0$$

$$\Gamma^3_{13} = \Gamma^3_{31} = g^{3j}\Gamma_{j,13} = g^{33}\Gamma_{3,13} = +\frac{1}{r}$$

$$\Gamma^3_{22} = g^{3j}\Gamma_{j,22} = g^{33}\Gamma_{3,22} = 0$$

$$\Gamma^3_{23} = \Gamma^3_{32} = g^{3j}\Gamma_{j,23} = g^{33}\Gamma_{3,23} = \frac{\cos\theta}{\sin\theta} = \cot\theta$$

$$\Gamma^3_{33} = g^{3j}\Gamma_{j,33} = g^{33}\Gamma_{3,33} = 0 \qquad\qquad (22.10)$$

From the results listed in (22.10) we can make the following conclusions:

$$\Gamma^k_{00} \equiv 0, \quad \Gamma^0_{0k} = \Gamma^0_{k0} \equiv 0, \tag{22.11}$$

and

$$\Gamma^0_{\alpha\beta} = -\dot{R}g_{\alpha\beta}, \quad \Gamma^\beta_{\alpha 0} = \Gamma^\beta_{0\alpha} = \delta^\beta_\alpha \frac{\dot{R}}{R}, \tag{22.12}$$

where $(\alpha, \beta = 1, 2, 3)$. Thus we further obtain

$$\Gamma^0_{\alpha\beta} \equiv 0, \quad \Gamma^1_{\alpha\beta} \equiv 0 \quad \text{for } \alpha \neq \beta,$$

$$\Gamma^2_{\alpha\beta} = \delta^1_\alpha \delta^2_\beta \frac{1}{r}, \quad \Gamma^3_{\alpha\beta} = \delta^1_\alpha \delta^3_\beta \frac{1}{r} + \delta^2_\alpha \delta^3_\beta \cot\theta \quad \text{for } \alpha \neq \beta. \tag{22.13}$$

Using the results (22.12) and (22.13) we note that

$$\partial_j \Gamma^j_{\alpha\beta} = 0 \quad \text{for } \alpha \neq \beta, \quad \partial_\beta \Gamma^\beta_{\alpha 0} = 0. \tag{22.14}$$

Furthermore, we can calculate the following sums:

$$\Gamma^j_{0j} = \Gamma^0_{00} + \Gamma^1_{01} + \Gamma^2_{02} + \Gamma^3_{03} = +3\frac{\dot{R}}{R}$$

$$\Gamma^j_{1j} = \Gamma^0_{10} + \Gamma^1_{11} + \Gamma^2_{12} + \Gamma^3_{13} = \frac{kr}{1 - kr^2} + \frac{2}{r}$$

$$\Gamma^j_{2j} = \Gamma^0_{20} + \Gamma^1_{21} + \Gamma^2_{22} + \Gamma^3_{23} = \cot\theta$$

$$\Gamma^j_{3j} = \Gamma^0_{30} + \Gamma^1_{31} + \Gamma^2_{32} + \Gamma^3_{33} = 0. \tag{22.15}$$

Using the results (22.15) we note that

$$\partial_\beta \Gamma^j_{\alpha j} = 0 \quad \text{for } \alpha \neq \beta, \quad \partial_0 \Gamma^j_{\alpha j} = 0. \tag{22.16}$$

The Ricci tensor for the metric (22.1) can be calculated using the definition (18.8), i.e.,

$$R_{kn} = \partial_n \Gamma^j_{kj} - \partial_j \Gamma^j_{kn} + \Gamma^p_{kj}\Gamma^j_{pn} - \Gamma^p_{kn}\Gamma^j_{pj}. \tag{22.17}$$

From the result (12.48) we see that the Ricci tensor is a symmetric tensor $R_{kn} = R_{nk}$ and that it has only 10 independent components. The formulae for the components of the Ricci tensor for the metric (22.1) are summarized in the following list:

$$R_{00} = \partial_0 \Gamma^j_{0j} - \partial_j \Gamma^j_{00} + \Gamma^p_{0j}\Gamma^j_{p0} - \Gamma^p_{00}\Gamma^j_{pj}$$

$$R_{01} = R_{10} = \partial_r \Gamma^j_{0j} - \partial_j \Gamma^j_{01} + \Gamma^p_{0j}\Gamma^j_{p1} - \Gamma^p_{01}\Gamma^j_{pj}$$

$$R_{02} = R_{20} = \partial_\theta \Gamma^j_{0j} - \partial_j \Gamma^j_{02} + \Gamma^p_{0j}\Gamma^j_{p2} - \Gamma^p_{02}\Gamma^j_{pj}$$

$$R_{03} = R_{30} = \partial_\varphi \Gamma^j_{0j} - \partial_j \Gamma^j_{03} + \Gamma^p_{0j}\Gamma^j_{p3} - \Gamma^p_{03}\Gamma^j_{pj}$$

$$R_{11} = \partial_r \Gamma^j_{1j} - \partial_j \Gamma^j_{11} + \Gamma^p_{1j}\Gamma^j_{p1} - \Gamma^p_{11}\Gamma^j_{pj}$$

$$R_{12} = R_{21} = \partial_\theta \Gamma^j_{1j} - \partial_j \Gamma^j_{12} + \Gamma^p_{1j}\Gamma^j_{p2} - \Gamma^p_{12}\Gamma^j_{pj}$$

$$R_{13} = R_{31} = \partial_\varphi \Gamma^j_{1j} - \partial_j \Gamma^j_{13} + \Gamma^p_{1j}\Gamma^j_{p3} - \Gamma^p_{13}\Gamma^j_{pj}$$

$$R_{22} = \partial_\theta \Gamma^j_{2j} - \partial_j \Gamma^j_{22} + \Gamma^p_{2j}\Gamma^j_{p2} - \Gamma^p_{22}\Gamma^j_{pj}$$

$$R_{23} = R_{32} = \partial_\varphi \Gamma^j_{2j} - \partial_j \Gamma^j_{23} + \Gamma^p_{2j}\Gamma^j_{p3} - \Gamma^p_{23}\Gamma^j_{pj}$$

$$R_{33} = \partial_\varphi \Gamma^j_{3j} - \partial_j \Gamma^j_{33} + \Gamma^p_{3j}\Gamma^j_{p3} - \Gamma^p_{33}\Gamma^j_{pj}. \tag{22.18}$$

The results for all 10 individual components of the Ricci tensor for the metric (22.1) can be obtained from this list (22.18). However, the calculation process can be facilitated using the general results (22.11)–(22.16). Thus for $\alpha, \beta = 1, 2, 3$, and $\alpha \neq \beta$, we may calculate

$$R_{\alpha\beta} = \partial_\beta \Gamma^j_{\alpha j} - \partial_j \Gamma^j_{\alpha\beta} + \Gamma^p_{\alpha j}\Gamma^j_{p\beta} - \Gamma^p_{\alpha\beta}\Gamma^j_{pj}. \tag{22.19}$$

From (22.14) and (22.16) we see that the first two terms on the right-hand side of Equation (22.19) for $\alpha \neq \beta$ vanish and we may write

$$R_{\alpha\beta} = \Gamma^p_{\alpha j}\Gamma^j_{p\beta} - \Gamma^0_{\alpha\beta}\Gamma^j_{0j} - \Gamma^1_{\alpha\beta}\Gamma^j_{1j} - \Gamma^2_{\alpha\beta}\Gamma^j_{2j}. \tag{22.20}$$

Using here $\Gamma^0_{\alpha\beta} = 0$ and $\Gamma^1_{\alpha\beta} = 0$ for $\alpha \neq \beta$ from (22.13), we obtain

$$R_{\alpha\beta} = \Gamma^p_{\alpha j}\Gamma^j_{p\beta} - \delta^1_\alpha \delta^2_\beta \frac{\cot\theta}{r} \qquad \text{for } \alpha \neq \beta. \tag{22.21}$$

Thus the second term on the right-hand side of Equation (22.21) is not equal to zero only for $(\alpha\beta) = (12)$. Let us now calculate the first term on the right-hand side of Equation (22.21), as follows:

$$\Gamma^p_{\alpha j}\Gamma^j_{p\beta} = \Gamma^p_{\alpha 0}\Gamma^0_{p\beta} + \Gamma^p_{\alpha\omega}\Gamma^\omega_{p\beta} \tag{22.22}$$

or

$$\Gamma^p_{\alpha j}\Gamma^j_{p\beta} = \Gamma^0_{\alpha 0}\Gamma^0_{0\beta} + \Gamma^\sigma_{\alpha 0}\Gamma^0_{\sigma\beta} + \Gamma^0_{\alpha\omega}\Gamma^\omega_{0\beta} + \Gamma^\sigma_{\alpha\omega}\Gamma^\omega_{\sigma\beta}. \tag{22.23}$$

It is shown by direct calculation that the first three terms on the right-hand side of Equation (22.23) vanish. Thus we obtain

$$\Gamma^p_{\alpha j}\Gamma^j_{p\beta} = \Gamma^\sigma_{\alpha\omega}\Gamma^\omega_{\sigma\beta} = \Gamma^1_{\alpha\omega}\Gamma^\omega_{1\beta} + \Gamma^2_{\alpha\omega}\Gamma^\omega_{2\beta} + \Gamma^3_{\alpha\omega}\Gamma^\omega_{3\beta}$$

$$= \Gamma^1_{\alpha 1}\Gamma^1_{1\beta} + \Gamma^1_{\alpha 2}\Gamma^2_{1\beta} + \Gamma^1_{\alpha 3}\Gamma^3_{1\beta} + \Gamma^2_{\alpha 1}\Gamma^1_{2\beta} + \Gamma^2_{\alpha 2}\Gamma^2_{2\beta} + \Gamma^2_{\alpha 3}\Gamma^3_{2\beta}$$

$$+ \Gamma^3_{\alpha 1}\Gamma^1_{3\beta} + \Gamma^3_{\alpha 2}\Gamma^2_{3\beta} + \Gamma^3_{\alpha 3}\Gamma^3_{3\beta}. \tag{22.24}$$

Using the result (22.24) we can show by direct calculation that

$$\Gamma^p_{\alpha j}\Gamma^j_{p\beta} = \delta^1_\alpha\delta^2_\beta\Gamma^3_{13}\Gamma^3_{32} = \delta^1_\alpha\delta^2_\beta\frac{\cot\theta}{r}. \tag{22.25}$$

Substituting (22.25) into (22.21), we finally obtain

$$R_{\alpha\beta} = R_{\beta\alpha} \equiv 0 \quad (\alpha \neq \beta). \tag{22.26}$$

Furthermore for $\alpha = 1, 2, 3$, we may also calculate

$$R_{\alpha 0} = \partial_0\Gamma^j_{\alpha j} - \partial_0\Gamma^0_{\alpha 0} - \partial_\beta\Gamma^\beta_{\alpha 0} + \Gamma^p_{\alpha j}\Gamma^j_{p0} - \Gamma^p_{\alpha 0}\Gamma^j_{pj}. \tag{22.27}$$

Using here the results (22.11), (22.14), and (22.16), we see that the first three terms on the right-hand side of Equation (22.27) vanish. We then have

$$R_{\alpha 0} = \Gamma^p_{\alpha 0}\Gamma^0_{p0} + \Gamma^p_{\alpha\beta}\Gamma^\beta_{p0} - \Gamma^0_{\alpha 0}\Gamma^j_{0j} - \Gamma^\beta_{\alpha 0}\Gamma^j_{\beta j}. \tag{22.28}$$

Using (22.11) further, the first and third terms on the right-hand side of Equation (22.28) vanish. Thus we obtain

$$R_{\alpha 0} = \Gamma^0_{\alpha\beta}\Gamma^\beta_{00} + \Gamma^\nu_{\alpha\beta}\Gamma^\beta_{\nu 0} - \Gamma^\beta_{\alpha 0}\Gamma^0_{\beta 0} - \Gamma^\beta_{\alpha 0}\Gamma^\nu_{\beta\nu}. \tag{22.29}$$

Using once again the results (22.11), the first and third terms on the right-hand side of Equation (22.29) vanish. We may then write

$$R_{\alpha 0} = \Gamma^\nu_{\alpha\beta}\Gamma^\beta_{\nu 0} - \Gamma^\beta_{\alpha 0}\Gamma^\nu_{\beta\nu}. \tag{22.30}$$

Using the result (22.24) we can show by direct calculation that

$$R_{\alpha 0} \equiv 0 \quad (\alpha = 1, 2, 3). \tag{22.31}$$

From the results (22.26) and (22.31) we conclude that all off-diagonal components of the Ricci tensor are identically equal to zero:

$$R_{kn} = R_{nk} \equiv 0 \quad (k \neq n). \tag{22.32}$$

Thus the only nontrivial components of the Ricci tensor are the diagonal components. Using the results (22.11)–(22.16), they can be calculated as

$$R_{00} = \partial_0 \Gamma^j_{0j} + \Gamma^p_{0j} \Gamma^j_{p0}$$

$$= 3\partial_0 \left(\frac{\dot{R}}{R} \right) + \left(\Gamma^1_{01} \right)^2 + \left(\Gamma^2_{02} \right)^2 + \left(\Gamma^3_{03} \right)^2 = 3\frac{\ddot{R}}{R}$$

$$R_{11} = \partial_r \Gamma^j_{1j} - \partial_0 \Gamma^0_{11} - \partial_r \Gamma^1_{11} + \Gamma^p_{1j} \Gamma^j_{p1} - \Gamma^0_{11} \Gamma^j_{0j} - \Gamma^1_{11} \Gamma^j_{1j}$$

$$= \partial_r \left(\frac{kr}{1 - kr^2} + \frac{2}{r} \right) - \partial_0 \left(\frac{\dot{R}R}{1 - kr^2} \right) - \partial_r \left(\frac{kr}{1 - kr^2} \right)$$

$$+ \frac{2\dot{R}^2}{1 - kr^2} + \frac{k^2 r^2}{(1 - kr^2)^2} + \frac{2}{r^2} - \frac{3\dot{R}^2}{1 - kr^2}$$

$$- \frac{kr}{1 - kr^2} \left(\frac{kr}{1 - kr^2} + \frac{2}{r} \right) = -\frac{R^2}{1 - kr^2} \left(\frac{\ddot{R}}{R} + \frac{2\dot{R}^2 + 2k}{R^2} \right)$$

$$R_{22} = \partial_\theta \Gamma^j_{2j} - \partial_0 \Gamma^0_{22} - \partial_r \Gamma^1_{22} + \Gamma^p_{2j} \Gamma^j_{p2} - \Gamma^0_{22} \Gamma^j_{0j} - \Gamma^1_{22} \Gamma^j_{1j}$$

$$= \partial_\theta (\cot \theta) - \partial_0 \left(R\dot{R}r^2 \right) - \partial_r \left[-r \left(1 - kr^2 \right) \right] + 2\dot{R}^2 r^2$$

$$- 2 \left(1 - kr^2 \right) + \cot^2 \theta - 3\dot{R}^2 r^2 + 2 - kr^2$$

$$= -R^2 r^2 \left(\frac{\ddot{R}}{R} + \frac{2\dot{R}^2 + 2k}{R^2} \right)$$

$$R_{33} = -\partial_0 \Gamma^0_{33} - \partial_r \Gamma^1_{33} - \partial_\theta \Gamma^2_{33} + \Gamma^p_{3j} \Gamma^j_{p3} - \Gamma^0_{33} \Gamma^j_{0j} - \Gamma^1_{33} \Gamma^j_{1j}$$

$$- \Gamma^2_{33} \Gamma^j_{2j} = -\partial_0 \left(R\dot{R}r^2 \sin^2 \theta \right) - \partial_r \left[-r \left(1 - kr^2 \right) \sin^2 \theta \right]$$

$$- \partial_\theta (-\sin \theta \cos \theta) + 2\dot{R}^2 r^2 \sin^2 \theta - 2 \left(1 - kr^2 \right) \sin^2 \theta$$

$$- 2 \cos^2 \theta - 3\dot{R}^2 r^2 \sin^2 \theta + \left(2 - kr^2 \right) \sin^2 \theta + \cos^2 \theta$$

$$= -R^2 r^2 \sin^2 \theta \left(\frac{\ddot{R}}{R} + \frac{2\dot{R}^2 + 2k}{R^2} \right). \tag{22.33}$$

Multiplying the covariant components of the Ricci tensor (22.33) by the corresponding components of the contravariant metric tensor given by (22.3), we obtain the components of the mixed Ricci tensor as follows:

$$R_0^0 = g^{00} R_{00} = 3\frac{\ddot{R}}{R}$$

$$R_1^1 = g^{11} R_{11} = \frac{\ddot{R}}{R} + \frac{2\dot{R}^2 + 2k}{R^2}$$

$$R_2^2 = g^{22} R_{22} = \frac{\ddot{R}}{R} + \frac{2\dot{R}^2 + 2k}{R^2}$$

$$R_3^3 = g^{33} R_{33} = \frac{\ddot{R}}{R} + \frac{2\dot{R}^2 + 2k}{R^2}. \tag{22.34}$$

The Ricci scalar R_j^j is then given by

$$R_j^j = 3\frac{\ddot{R}}{R} + 3\left(\frac{\ddot{R}}{R} + \frac{2\dot{R}^2 + 2k}{R^2}\right) = 6\left(\frac{\ddot{R}}{R} + \frac{\dot{R}^2 + k}{R^2}\right) \tag{22.35}$$

Let us now form a mixed tensor G_n^k, entering the gravitational field Equations (18.97) as

$$G_n^k = R_n^k - \frac{1}{2}\delta_n^k R_j^j = -\frac{8\pi G}{c^4} T_n^k. \tag{22.36}$$

The tensor G_n^k is usually called the *Einstein tensor*. Substituting the results (22.34) and (22.35) into the definition (22.36), we obtain the components of the Einstein tensor for the Robertson–Walker metric as follows:

$$G_0^0 = R_0^0 - \frac{1}{2}R_j^j = 3\frac{\ddot{R}}{R} - 3\left(\frac{\ddot{R}}{R} + \frac{\dot{R}^2 + k}{R^2}\right) = -3\frac{\dot{R}^2 + k}{R^2}$$

$$G_1^1 = G_2^2 = G_3^3 = R_1^1 - \frac{1}{2}R_j^j = \frac{\ddot{R}}{R} + \frac{2\dot{R}^2 + 2k}{R^2}$$

$$- 3\frac{\ddot{R}}{R} - 3\frac{\dot{R}^2 + k}{R^2} = -\left(2\frac{\ddot{R}}{R} + \frac{\dot{R}^2 + k}{R^2}\right). \tag{22.37}$$

At this stage it is appropriate to reverse the temporary convention (22.6) and to return to the usual definition of dots as the derivatives of the scale radius $R(t)$ with respect to the cosmic time t, i.e.,

$$\dot{R}(t) = \frac{dR(t)}{dt}. \tag{22.38}$$

This change of convention introduces an additional factor of $1/c$ before each derivative defined by the dots. Thus the results (22.37) become

$$G_0^0 = -\frac{3}{c^2}\frac{\dot{R}^2 + kc^2}{R^2}$$

$$G_1^1 = G_2^2 = G_3^3 = -\frac{1}{c^2}\left(2\frac{\ddot{R}}{R} + \frac{\dot{R}^2 + kc^2}{R^2}\right). \qquad (22.39)$$

Equations (22.39) are the final results for the components of the Einstein tensor in the Robertson–Walker metric.

22.2 The Friedmann Equations

In order to construct the gravitational field Equations (22.36), we now need the energy–momentum tensor T_n^k of the cosmic matter. The assumption is that the matter in the universe is, on the large scale, evenly distributed in the form of a cosmic fluid with no shear-viscous, bulk-viscous, or heat-conductive features. Thus we may use the mixed energy–momentum tensor of an ideal fluid, which is in the covariant form given by (18.72). We may then write

$$T_n^k = \left(p + \rho c^2\right)u_n u^k - \delta_n^k p, \qquad (22.40)$$

where p is the pressure and ρ is the matter (or rest energy) density of the cosmic fluid. In the co-moving frame the cosmic fluid is at rest and we have

$$u^0 = 1, \quad u^\alpha = 0 \quad (\alpha = 1, 2, 3). \qquad (22.41)$$

Using the results (22.41) and observing that $p \ll \rho c^2$, we obtain

$$T_0^0 = \rho c^2, \quad T_1^1 = T_2^2 = T_3^3 = -p. \qquad (22.42)$$

Thus the energy–momentum tensor of the cosmic fluid in the co-moving frame can be structured in the following matrix:

$$[T_n^k] = \begin{bmatrix} \rho c^2 & 0 & 0 & 0 \\ 0 & -p & 0 & 0 \\ 0 & 0 & -p & 0 \\ 0 & 0 & 0 & -p \end{bmatrix}. \qquad (22.43)$$

Substituting the results (22.39) and (22.42) into (22.36), we obtain the gravitational field equations in the form

$$\frac{\dot{R}^2 + kc^2}{R^2} = \frac{8\pi G}{3}\rho \tag{22.44}$$

$$2\frac{\ddot{R}}{R} + \frac{\dot{R}^2 + kc^2}{R^2} = -\frac{8\pi G}{c^2}p. \tag{22.45}$$

Equations (22.44) and (22.45) are called the *Friedmann equations*, and their solution describes cosmic dynamics. Combining Equations (22.44) and (22.45), we may write

$$2\frac{\ddot{R}}{R} + \frac{8\pi G}{3}\rho = -\frac{8\pi G}{c^2}p. \tag{22.46}$$

Furthermore, from Equation (22.44) we have

$$\dot{R}^2 + kc^2 = \frac{8\pi G}{3}\rho R^2. \tag{22.47}$$

Differentiating Equation (22.47) with respect to the cosmic time t, we obtain

$$2\dot{R}\ddot{R} = \frac{8\pi G}{3}\dot{\rho}R^2 + \frac{8\pi G}{3}\rho 2R\dot{R}, \tag{22.48}$$

or

$$2\frac{\ddot{R}}{R} = \frac{8\pi G}{3}\dot{\rho}\frac{R}{\dot{R}} + \frac{8\pi G}{3}2\rho. \tag{22.49}$$

Substituting Equation (22.49) into (22.46), we obtain

$$\frac{8\pi G}{3}\dot{\rho}\frac{R}{\dot{R}} + \frac{8\pi G}{3}3\rho = -\frac{8\pi G}{3}3\frac{p}{c^2} \tag{22.50}$$

or

$$\dot{\rho}\frac{R}{\dot{R}} + 3\left(\rho + \frac{p}{c^2}\right) = 0. \tag{22.51}$$

Observing that the universe as a system of galaxies in the smooth fluid approximation behaves as an incoherent dust, we can again use the approximation $p \ll \rho c^2$ to obtain

$$\dot{\rho} + 3\rho\frac{\dot{R}}{R} = \frac{1}{R^3}\frac{d}{dt}\left(\rho R^3\right) = 0. \tag{22.52}$$

Integrating (22.52) we obtain

$$\rho R^3 = \rho_0 R_0^3 = \text{Constant},\qquad\qquad (22.53)$$

where ρ_0 and R_0 are the matter density and scale radius at the present epoch ($t = t_0$). From the result (22.53) we see that the matter density varies in time as R^{-3} and that the quantity of matter in a co-moving volume element is constant during the expansion of the universe. This conclusion is relevant to the present matter-dominated epoch in the history of the universe.

   In the early phases of its history when the universe was radiation dominated, this conclusion was not valid because at that stage the assumption $p \ll \rho c^2$ was not valid either. In the radiation-dominated universe, Equation (22.51) becomes

$$\dot{\rho} + 3\left(\rho + \frac{p}{c^2}\right)\frac{\dot{R}}{R} = 0,\qquad\qquad (22.54)$$

and the pressure component cannot be neglected. Using (22.53) and neglecting the pressure, we obtain the Friedmann equations for the matter-dominated epoch of the universe as follows:

$$2\frac{\ddot{R}}{R} + \frac{\dot{R}^2 + kc^2}{R^2} = 0\qquad\qquad (22.55)$$

$$\frac{\dot{R}^2 + kc^2}{R^2} = \frac{8\pi G\rho_0}{3}\frac{R_0^3}{R^3}.\qquad\qquad (22.56)$$

The solutions of the Friedmann equations for the matter-dominated epoch of the universe given by (22.55) and (22.56) will be the subject of the next chapter.

Chapter 23

Nonstatic Models of the Universe

In the previous chapter we derived the cosmic dynamic equations satisfied by the scale radius $R(t)$, known as the Friedmann equations. As we concluded earlier, the nonstatic models of the universe based on the Robertson–Walker metric (21.49) are described by their scale radius $R(t)$ and the curvature constant k. The objective of the present chapter is therefore to study the appropriate solutions for the scale radius $R(t)$ and the appropriate values of the curvature constant k and their implications for the future evolution and other physical properties of the expanding universe.

23.1 | Solutions of the Friedmann Equations

In order to solve the Friedmann equations (22.55) and (22.56), we first introduce some useful quantities. Let us recall that the Hubble constant (21.54) is given by

$$H(t) = \frac{\dot{R}(t)}{R(t)}.$$

(23.1)

Substituting (23.1) into (22.56) we obtain

$$H^2 + \frac{kc^2}{R^2} = \frac{8\pi G\rho_0}{3}\frac{R_0^3}{R^3}.$$

(23.2)

In the present epoch ($t = t_0$) with $R = R_0$ and $H = H_0$, Equation (23.2) becomes

$$H_0^2 + \frac{kc^2}{R_0^2} = \frac{8\pi G\rho_0}{3}$$

(23.3)

or

$$\frac{k}{R_0^2} = \frac{8\pi G\rho_0}{3c^2} - \frac{H_0^2}{c^2} = \frac{8\pi G}{3c^2}(\rho_0 - \rho_C),$$

(23.4)

where ρ_C is the critical (or closure) matter density of the universe, defined by

$$\rho_C = \frac{3H_0^2}{8\pi G}.$$

(23.5)

With the range of the estimated values of the Hubble constant H_0 in the present epoch, the numerical value of the critical density is given by

$$\rho_C = 2 \times 10^{-26} a^2 \frac{kg}{m^3} \quad (0.5 \le a \le 1).$$

(23.6)

Let us also recall the definition of the deceleration parameter $q(t)$ defined according to Equation (21.65) as follows:

$$q(t) = -\frac{\ddot{R}(t)R(t)}{\dot{R}^2(t)} = -\frac{1}{H^2}\frac{\ddot{R}}{R}.$$

(23.7)

From the Friedmann equations (22.55) and (22.56) we may write

$$\frac{\ddot{R}}{R} = -\frac{4\pi G\rho_0}{3}\frac{R_0^3}{R^3}.$$

(23.8)

Substituting (23.8) into (23.7) we obtain

$$q(t) = \frac{4\pi G\rho_0}{3H^2}\frac{R_0^3}{R^3}.$$

(23.9)

The present-epoch value of the deceleration parameter (23.9) is given by

$$q_0 = \frac{4\pi G\rho_0}{3H_0^2} = \frac{\rho_0}{2\rho_C}.$$

(23.10)

From (23.10) we see that the present-epoch value of the deceleration parameter can be expressed in terms of the present and critical matter densities of the universe.

23.1.1 The Flat Model ($k = 0$)

For the flat model with $k = 0$, Equation (23.4) gives $\rho_0 = \rho_C$ and from Equation (23.10) we find $q_0 = 1/2$. Using $\rho_0 = \rho_C$ and (23.5) we may write

$$H_0^2 = \frac{8\pi G \rho_0}{3}. \tag{23.11}$$

Friedmann Equation (22.56) then becomes

$$\dot{R}^2 = \frac{8\pi G \rho_0 R_0^3}{3R} = \frac{H_0^2 R_0^3}{R}, \tag{23.12}$$

or

$$\dot{R}^2 = \frac{A^2}{R}, \quad A = H_0 R_0^{3/2}. \tag{23.13}$$

Equation (23.13) can be rewritten as follows:

$$\dot{R} = \frac{dR}{dt} = AR^{-1/2} \implies \int \sqrt{R} dR = A \int dt + C. \tag{23.14}$$

Performing the elementary integration in (23.14) we obtain

$$\frac{2}{3} R^{3/2} = At + C \implies R(t) = \left(\frac{3A}{2} \right)^{2/3} t^{2/3}, \tag{23.15}$$

where we used the initial condition $R(0) = 0$ to set the integration constant C in (23.15) equal to zero. From Equation (23.15) we may also write

$$t = \frac{2}{3A} R^{3/2} = \frac{2}{3H_0} \left(\frac{R}{R_0} \right)^{3/2}. \tag{23.16}$$

The present-epoch Equation (23.16) for $t = t_0$ has the form

$$t_0 = \frac{2}{3H_0}. \tag{23.17}$$

The solution (23.15) for the scale radius $R(t)$ with the choice of the curvature constant $k = 0$, is, in the literature, sometimes called the *Einstein–de Sitter Solution*.

23.1.2 The Closed Model ($k = 1$)

For the closed model with $k = 1$, Equation (23.4) gives $\rho_0 > \rho_C$, and from Equation (23.10) we find $q_0 > 1/2$. The Friedmann Equation (22.56) then becomes

$$\frac{\dot{R}^2}{c^2} + 1 = \frac{8\pi G\rho_0 R_0^3}{3c^2 R} = \frac{B}{R}, \tag{23.18}$$

where we introduce a constant B as

$$B = \frac{8\pi G\rho_0 R_0^3}{3c^2} = 2\frac{4\pi G\rho_0}{3H_0^2}\frac{H_0^2 R_0^3}{c^2}. \tag{23.19}$$

Using the definition of the present-epoch deceleration parameter (23.10) we obtain

$$B = 2q_0 \frac{H_0^2 R_0^3}{c^2}. \tag{23.20}$$

Furthermore, using Equation (23.4) for $k = 1$, we may write

$$\frac{1}{R_0^2} = \frac{H_0^2}{c^2}\left(2\frac{4\pi G\rho_0}{3H_0^2} - 1\right) = \frac{H_0^2}{c^2}(2q_0 - 1), \tag{23.21}$$

or

$$R_0 = \frac{c}{H_0}(2q_0 - 1)^{-1/2}. \tag{23.22}$$

Substituting (23.22) into (23.20) we obtain

$$B = \frac{2q_0}{(2q_0 - 1)^{3/2}}\frac{c}{H_0}. \tag{23.23}$$

Equation (23.18) may be rewritten in the form

$$\frac{1}{c}\frac{dR}{dt} = \frac{B - R}{R} \implies \int_0^t c\,dt = \int_0^R \sqrt{\frac{R}{B - R}}\,dR. \tag{23.24}$$

Let us now introduce an angular parameter η as follows:

$$R = B \sin^2 \frac{\eta}{2} = \frac{B}{2}(1 - \cos \eta), \tag{23.25}$$

with

$$dR = B \sin\frac{\eta}{2}\cos\frac{\eta}{2}d\eta. \tag{23.26}$$

Using Equations (23.25) and (23.26) we may write

$$\sqrt{\frac{R}{B - R}}\,dR = B \sin^2\frac{\eta}{2}d\eta = \frac{B}{2}(1 - \cos \eta)\,d\eta. \tag{23.27}$$

Using (23.25) and performing the elementary integration in (23.24) with (23.27), we obtain the two parameter equations

$$R = \frac{B}{2}(1 - \cos \eta), \quad ct = \frac{B}{2}(\eta - \sin \eta). \quad (23.28)$$

The parameter Equations (23.28) define the scale radius $R(t)$ by a parameter η. The function $R(t)$ defined by (23.28) is a cycloid, and the scale radius is a cyclic function of the cosmic time. The universe starts expansion at the cosmic time $t = 0$ ($\eta = 0$) with scale radius $R = 0$ and expands to a maximum size $R = B$ at the cosmic time $t = t_L/2$ ($\eta = \pi$). Thereafter the universe contracts back to the scale radius $R = 0$ at the cosmic time $t = t_L$ ($\eta = 2\pi$). The time t_L in the closed model is called the *lifespan* of the present universe. It can be calculated from (23.28) with (23.23) for $\eta = 2\pi$, as follows:

$$t_L = \frac{B}{2c}2\pi = \frac{\pi B}{c} = \frac{2\pi q_0}{(2q_0 - 1)^{3/2}} \frac{1}{H_0}. \quad (23.29)$$

For example, the choice $q_0 = 1$ gives a lifespan of the universe equal to $t_L = 2\pi/H_0$. From Equations (23.22) and (23.23) we also note that for $q_0 = 1$, the present scale radius $R_0 = c/H_0$ is equal to one half of the maximum scale radius $B = 2c/H_0$. It means that in the closed model for $q_0 = 1$, the present universe will expand to twice its present size before it starts contracting.

23.1.3 The Open Model ($k = -1$)

For the open model with $k = -1$, Equation (23.4) gives $\rho_0 < \rho_C$, and from Equation (23.10) we find $q_0 < 1/2$. The Friedmann Equation (22.56) then becomes

$$\frac{\dot{R}^2}{c^2} - 1 = \frac{8\pi G \rho_0 R_0^3}{3c^2 R} = \frac{B}{R}, \quad (23.30)$$

where we again use the constant B given by (23.23). Equation (23.30) may be rewritten in the form

$$\frac{1}{c}\frac{dR}{dt} = \frac{B + R}{R} \Rightarrow \int_0^t c\,dt = \int_0^R \sqrt{\frac{R}{B + R}}\,dR. \quad (23.31)$$

Let us now introduce an angular parameter η as follows:

$$R = B \sinh^2 \frac{\eta}{2} = \frac{B}{2}(\cosh \eta - 1), \quad (23.32)$$

with

$$dR = B \sinh \frac{\eta}{2} \cosh \frac{\eta}{2} d\eta. \tag{23.33}$$

Using Equations (23.32) and (23.33), we may write

$$\sqrt{\frac{R}{B+R}} dR = B \sinh^2 \frac{\eta}{2} d\eta = \frac{B}{2} (\cosh \eta - 1) \, d\eta. \tag{23.34}$$

Using (23.32) and performing elementary integrations in (23.31) with (23.34), we obtain the two parameter equations

$$R = \frac{B}{2} (\cosh \eta - 1), \quad ct = \frac{B}{2} (\sinh \eta - \eta). \tag{23.35}$$

The parameter Equations (23.35) define the scale radius $R(t)$ by a parameter η. The function $R(t)$ defined by (23.35) is growing indefinitely as $t \to \infty$ ($\eta \to \infty$). Like the flat Einstein–de Sitter solution, the open solution continues to expand forever. It should also be noted that the expansion of the universe in the open model is faster than that of the flat model because of the presence of the exponential functions in the parameter η.

23.2 Closed or Open Universe

The analysis of the three possible models for the expansion of the universe, presented in the previous chapter, does not resolve the question whether the present universe is open or closed. The answer to that question must be looked for in astronomical observations and estimates of the various parameters of the model. The current estimates of the present-epoch Hubble constant H_0 and the critical density ρ_C are

$$H_0 = 100a \frac{\text{km}}{\text{s}} \frac{1}{\text{Mpc}}$$

$$\rho_C = 2 \times 10^{-26} a^2 \frac{\text{kg}}{\text{m}^3}, \quad 0.5 \leq a \leq 1. \tag{23.36}$$

Estimates of the present-epoch deceleration parameter q_0 are more difficult to obtain. In order to study the deceleration parameter q_0, we use the result for the Hubble constant $H = \dot{R}/R$ to calculate the second derivative of the scale radius with respect to cosmic time as follows:

$$\dot{R} = HR \implies \ddot{R} = H\dot{R} + \dot{H}R = H^2 R + \dot{H}R. \tag{23.37}$$

Substituting (23.37) into (23.7) we obtain

$$q(t) = -\frac{1}{H^2}\frac{\ddot{R}}{R} = -\left(\frac{\dot{H}}{H^2} + 1\right). \qquad (23.38)$$

Using (23.38) we can calculate the present-epoch deceleration parameter q_0 as

$$q_0 = -\left(\frac{\dot{H}(t_0)}{H^2(t_0)} + 1\right). \qquad (23.39)$$

The result (23.39) indicates that if we can measure $\dot{H}(t_0)$ and $H(t_0)$ we can calculate the present-epoch deceleration parameter q_0. In order to measure $\dot{H}(t_0)$ we use the fact that as we look deeper into space, we look farther back into time. For example, if we estimate the Hubble constant $H(t)$ for objects 1 billion light years away, we are really estimating the Hubble constant $H(t)$ for the universe 1 billion years back in the cosmic time. The difficulty with such a method of determining the present-epoch deceleration parameter q_0 is related to the difficulties of measuring the distances to the objects deep in space.

In spite of these difficulties, some existing observational data suggest the value of the present-epoch deceleration parameter $q_0 = 1$. As we have concluded before, for $q_0 > 1/2$, the universe is closed and such observational data suggest that the present universe is closed. However, using Equation (23.10) for $q_0 = 1$, we obtain the following present-epoch matter (or rest-energy) density of the universe:

$$\rho_0 = 2\rho_C = 4 \times 10^{-26}a^2\frac{\text{kg}}{\text{m}^3} \quad (0.5 \le a \le 1), \qquad (23.40)$$

which is much larger than the observed density of matter in the universe. This discrepancy is the origin of the so-called *problem of missing mass* in the universe. The problem was identified by measurements of the masses of clusters of galaxies in two different ways. One method was to exploit the definite relation between the luminosity of a galaxy and its mass, to sum up all the masses of member galaxies to obtain the total mass of a cluster. The other method was to measure the relative velocities between the member galaxies and to calculate the mean relative velocity, which is determined by the mass of the cluster. It was discovered that the luminosity mass obtained by the first method was considerably smaller than the dynamical mass obtained by the second method.

Thus there is reason to believe that there is a large amount of invisible matter within the observed clusters of galaxies. The first decisive evidence of the existence of such *dark matter* in the universe came from the rotation curves of galaxies. By measurements of the rotation curves of smaller galaxies revolving around great spiral galaxies, it was discovered that they differ completely from curves expected from Newtonian mechanics in empty space. The only possible interpretation of this result is that the space around galaxies is not empty. On the contrary, it seems to consist of a considerable amount of dark matter. It is today generally believed that up to 90 percent of the matter in the universe is such invisible dark matter. Thus we have the problem of determining the nature of the dark matter in the universe. Optical measurements indicate that the density of diffuse gas in the universe is far below the density required to account for the large amount of dark matter in the universe. Such gas could be in the form of high-temperature ionized gas emitting X-rays, and X-rays have indeed been detected in clusters of galaxies. However, the observed density is again far below the density required to account for the large amount of dark matter in the universe. There are enormous numbers of neutrinos in the present universe. Some experiments indicate that they may have a rest mass of the order of 10^{-35} kg, and if this is true the neutrinos may just be the missing matter. This is a growing area for combined efforts in astrophysics and particle physics today, although it is still not clear how the neutrinos could affect the rotation curves of galaxies in the observed way.

23.3 Newtonian Cosmology

In the present section we study the evolution of an expanding universe within the framework of the Newtonian theory of gravitation and compare the results with those obtained using the general theory of relativity. Let us imagine the large-scale matter-dominated Newtonian universe as an ideal fluid with negligible pressure moving according to the Newton laws of motion and gravity. Such fluid is often called the *Newtonian dust*. We now consider Newtonian dust with a uniform density $\rho(t)$ being in a state of uniform expansion. The only force acting on the dust particles is the force of gravity. The components of the three-dimensional position vector of an arbitrary dust particle at some cosmic time t are given by

$$x^{\alpha}(t) = R(t)\lambda^{\alpha} \quad (\alpha = 1, 2, 3), \qquad (23.41)$$

where λ is a constant vector defined by the initial position of the dust particle, and $R(t)$ is the scale radius of the uniform expansion. The first and second time derivatives of the coordinates (23.41) are given by

$$\dot{x}^{\alpha} = \dot{R}\lambda^{\alpha} = \frac{\dot{R}}{R}x^{\alpha} = Hx^{\alpha}, \quad \ddot{x}^{\alpha} = \ddot{R}\lambda^{\alpha} = \frac{\ddot{R}}{R}x^{\alpha}, \qquad (23.42)$$

where $H(t) = \dot{R}/R$ is the Hubble constant of the Newtonian expansion. Let us now use the continuity Equation (18.54) in the form

$$\frac{\partial \rho}{\partial t} + \partial_\alpha \left(\rho v^\alpha\right) = \frac{\partial \rho}{\partial t} + \dot{x}^\alpha \partial_\alpha \rho + \rho \partial_\alpha \dot{x}^\alpha = 0. \tag{23.43}$$

Using (23.42) and the definition of the total time derivative of the density ρ, given by

$$\dot{\rho} = \frac{d\rho}{dt} = \frac{\partial \rho}{\partial t} + \frac{\partial \rho}{\partial x^\alpha} \frac{dx^\alpha}{dt}, \tag{23.44}$$

we may rewrite (23.43) as follows:

$$\dot{\rho} + \rho H \partial_\alpha x^\alpha = 0 \Rightarrow \dot{\rho} + 3\rho H = 0, \tag{23.45}$$

or

$$\dot{\rho} + 3\frac{\dot{R}}{R}\rho = \frac{1}{R^3}\frac{d}{dt}\left(\rho R^3\right) = 0. \tag{23.46}$$

Integrating Equation (23.46) we recover the result (22.53) obtained using the Friedmann Equations, i.e.,

$$\rho R^3 = \rho_0 R_0^3 = \text{Constant}, \tag{23.47}$$

where ρ_0 and R_0 are the matter density and the scale radius at the present epoch $(t = t_0)$. From the result (23.47) we see that in Newtonian cosmology the matter density also varies in time as R^{-3}. The isotropy and spherical symmetry of the universe require that any spherical volume evolve only under its own influence. If the observed spherical volume has radius r and mass $M(r) = (4\pi/3)r^3\rho$, the equation of motion of a dust particle somewhere on its surface is given by the Newtonian Equation (17.44), as follows:

$$\ddot{x}^\alpha = \frac{\ddot{R}}{R}x^\alpha = -\frac{GM(r)}{r^3}x^\alpha = -\frac{4\pi G\rho}{3}x^\alpha. \tag{23.48}$$

As Equation (23.48) is valid for an arbitrary x^α, we obtain

$$\ddot{R} = -\frac{4\pi G\rho}{3}R. \tag{23.49}$$

Now using the result (23.47), Equation (23.49) becomes

$$\ddot{R} = -\frac{4\pi G\rho_0 R_0^3}{3R^2}. \tag{23.50}$$

Multiplying both sides of Equation (23.50) by $\dot{R}$, we obtain

$$\dot{R}\ddot{R} + \frac{4\pi G\rho_0 R_0^3}{3}\frac{\dot{R}}{R^2} = 0, \tag{23.51}$$

or

$$\frac{1}{2}\frac{d\dot{R}^2}{dt} - \frac{4\pi G\rho_0 R_0^3}{3}\frac{d}{dt}\left(\frac{1}{R}\right) = 0. \tag{23.52}$$

Thus we may write

$$\frac{d}{dt}\left(\dot{R}^2 - \frac{8\pi G\rho_0 R_0^3}{3R}\right) = 0. \tag{23.53}$$

Integrating Equation (23.53) we obtain

$$\dot{R}^2 - \frac{8\pi G\rho_0 R_0^3}{3R} = -kc^2, \tag{23.54}$$

where kc^2 is the integration constant. After some restructuring of Equation (23.54) we recover the second Friedmann equation (22.56), i.e.,

$$\frac{\dot{R}^2 + kc^2}{R^2} = \frac{8\pi G\rho_0}{3}\frac{R_0^3}{R^3}. \tag{23.55}$$

Furthermore, using the result (23.54) rewritten as

$$\frac{4\pi G\rho_0 R_0^3}{3R^2} = \frac{1}{2}\frac{\dot{R}^2 + kc^2}{R}, \tag{23.56}$$

Equation (23.50) becomes

$$\ddot{R} = -\frac{1}{2}\frac{\dot{R}^2 + kc^2}{R}. \tag{23.57}$$

Dividing by R and restructuring Equation (23.57), we recover the first Friedmann equation (22.55), i.e.,

$$2\frac{\ddot{R}}{R} + \frac{\dot{R}^2 + kc^2}{R^2} = 0. \tag{23.58}$$

From the preceding discussion we conclude that Newtonian cosmology gives the same dynamic equations for the scale factor $R(t)$ of the expanding universe as the relativistic cosmology based on the Robertson–Walker metric. In Newtonian cosmology the parameter k is just an integration constant, whereas relativistic cosmology k is the curvature constant. In fact if $k \neq 0$ there is no loss of generality if we set $k = \pm 1$ in Newtonian cosmology either. The dynamic Equations (23.55) and (23.58) lead to exactly the same three models of the expanding universe with $k = -1, 0, 1$. It should, however, be kept in mind that the Newtonian cosmology is just a nonrelativistic limit of the standard relativistic cosmological model and that it cannot account for a number of important physical features of the expanding universe. In particular it cannot handle the radiation-dominated early phases of the expanding universe and incorporate the pressure of the cosmic fluid in the proper way.

► **Chapter 24**

Quantum Cosmology

The objective of the present chapter is to give a short introduction into the most profound speculations about the earliest moments of the present universe. In the solutions of Friedmann equations, presented in the previous chapter, we have generally used the initial condition $R(0) = 0$. In effect such an initial condition means that in the earliest moments of its evolution, the present universe may have been so compact that its size was comparable to the size of a single quantum particle. In such an early epoch the classical solutions of the Friedmann equations, derived in the previous chapter, are no longer adequate and we need a quantum theory of the very early universe, i.e., a *quantum cosmology*. Since quantum cosmology is a highly speculative theory in its early development phase, the presentation in this chapter will be limited to a few introductory topics. Furthermore, as quantum theory is outside the scope of the present book, a few elementary quantum-mechanical results needed in the present chapter will be introduced without a detailed derivation from the first principles.

24.1 Introduction

The nonstatic models of the expansion of the universe, discussed in the previous chapter, are based on a well-formulated gravitational field theory and available observational data. These models can be used to study the early moments of the universe down to the times $t < 10^{-2}$ s. The subject

265

of the present chapter is to investigate the so-called *Planck epoch*, i.e., the times $t \leq 10^{-43}$ s.

The approach pursued in the present chapter is to use the canonical quantization procedure to quantize the gravitational degrees of freedom and to derive the Klein–Gordon wave equation for the wave function of the universe governing both the matter fields and the space-time geometry. Such a wave equation is generally known as the *Wheeler–DeWitt equation*. The wave function of the universe in this approach is a function of all possible three-dimensional geometries and matter field configurations, known as *superspace*. It should be noted that the concept of superspace used in quantum cosmology has nothing to do with the concept of superspace of the supersymmetric quantum theories.

In order to reduce the problem to a manageable one, all but a finite number of degrees of freedom are frozen out, leaving us with a finite-dimensional superspace known as the *mini-superspace*. In the present chapter we consider a simple mini-superspace model in which the only remaining degrees of freedom are the scale radius $R(t)$ of a closed homogeneous and isotropic universe and a homogeneous massive scalar field ϕ. Furthermore all the degrees of freedom of the scalar field are also frozen out, and the only effect of the scalar field is to provide the vacuum energy density ρ_{vac}. Such a vacuum energy density contributes to a cosmological constant term defined by $\Lambda = 8\pi G \rho_{vac}$.

Using the mini-superspace model just described, we then calculate the semiclassical wave functions of the universe and the tunneling probability for the universe to make the transition from $R = 0$ to the transition point $R = R_0$, being the upper limit of the classically forbidden region $0 < R < R_0$.

24.2 The Wheeler–DeWitt Equation

Let us start the derivation of the Wheeler–DeWitt equation by recalling the Friedmann equations of cosmic dynamics, given by (22.44) and (22.45), i.e.,

$$\frac{\dot{R}^2 + kc^2}{R^2} = \frac{8\pi G}{3}\rho \tag{24.1}$$

$$2\frac{\ddot{R}}{R} + \frac{\dot{R}^2 + kc^2}{R^2} = -\frac{8\pi G}{c^2}p. \tag{24.2}$$

In the very early phases of its evolution, the universe was clearly not matter dominated. In fact, it is generally believed that in the Planck epoch it was not even radiation dominated, and the energy–momentum tensor was determined by the vacuum energy. In such a vacuum-dominated epoch the pressure of the extremely dense matter of the universe is given by $p = -\rho c^2$. Using Equation (22.54), we see that in the vacuum-dominated epoch we have

$$\dot{\rho} = 0 \Rightarrow \rho = \rho_{vac} = \text{Constant.} \tag{24.3}$$

Thus the Friedmann Equations (24.1) and (24.2) in the closed model $(k = +1)$ become

$$\frac{\dot{R}^2 + c^2}{R^2} = \frac{8\pi G}{3} \rho_{vac} = \frac{\Lambda}{3} \tag{24.4}$$

$$2\frac{\ddot{R}}{R} + \frac{\dot{R}^2 + c^2}{R^2} = 8\pi G\rho_{vac} = \Lambda. \tag{24.5}$$

Substituting Equation (24.4) into (24.5) we obtain

$$2\frac{\ddot{R}}{R} + \frac{\Lambda}{3} = \Lambda \Rightarrow \frac{\ddot{R}}{R} = \frac{\Lambda}{3}. \tag{24.6}$$

Equation (24.6) can be written as follows:

$$\ddot{R} - \frac{c^2}{R^0}R = 0 \quad \left(R_0 = c\sqrt{\frac{3}{\Lambda}}\right). \tag{24.7}$$

The classical solution of Equation (24.7) is given by

$$R(t) = R_0 \cosh\left(\frac{ct}{R_0}\right). \tag{24.8}$$

This solution describes a universe that was infinitely large $(R \to \infty)$ in the infinite past $(t \to -\infty)$. Then it contracted to a minimum size of R_0 at $t = 0$, after which it started expanding again to an infinite size in the infinite future $(t \to +\infty)$. From Equation (24.8) we see that the classical solution has a forbidden range for $0 < R < R_0$. Thus it is classically impossible for the universe to start from $R = 0$ and evolve into the preceding cosmic model. However, given the quantum-mechanical nature of the problem in the Planck epoch, it may be possible for the universe to make such a transition with some finite quantum-mechanical tunneling probability. In order to calculate the wave functions of a universe starting at $R = 0$ and tunneling into the region $R > R_0$ and the tunneling probability of such a transition, we now construct the action integral of the universe in the Planck epoch as follows:

$$I_G = -\frac{c^4}{16\pi G} \int_t dt \mathcal{L}\left[R(t), \dot{R}(t)\right] \int_V \sqrt{-g}\, drd\theta d\varphi, \tag{24.9}$$

where

$$\sqrt{-g}drd\theta d\varphi = R^3 \frac{r^2 dr}{\sqrt{1-r^2}} \sin\theta d\theta d\varphi. \qquad (24.10)$$

In Equation (24.9) we have used the result for the Ricci scalar (22.35) in the Robertson–Walker metric with $k = +1$, i.e.,

$$R^j_j = \frac{6}{c^2}\left(\frac{\ddot{R}}{R} + \frac{\dot{R}^2 + c^2}{R^2}\right), \qquad (24.11)$$

to conclude that the Lagrangian density $\mathcal{L}\left[R(t), \dot{R}(t)\right]$ in the present model is space independent. Thus we can integrate out the spatial part of the action integral (24.9), which is just a three-dimensional spherically symmetric volume in the Robertson–Walker curved space-time metric (21.49), to obtain

$$\int_V \sqrt{-g}drd\theta d\varphi = R^3 \int_0^1 \frac{r^2 dr}{\sqrt{1-r^2}} \int_0^\pi \sin\theta d\theta \int_0^{2\pi} d\varphi. \qquad (24.12)$$

Integrating the elementary integrals over θ and φ and introducing the new angular variable χ as

$$r = \sin\chi \Rightarrow dr = \cos\chi d\chi, \qquad (24.13)$$

the result (24.12) becomes

$$\int_V \sqrt{-g}drd\theta d\varphi = 4\pi R^3 \int_0^{\pi/2} \sin^2\chi d\chi = 2\pi^2 R^3. \qquad (24.14)$$

Substituting (24.14) into (24.9) we obtain

$$I_G = -\frac{\pi c^4}{8G} \int_t dt R^3(t)\mathcal{L}\left[R(t), \dot{R}(t)\right]. \qquad (24.15)$$

Using the result for the Ricci scalar (24.11) in the Robertson–Walker metric with $k = +1$, we construct the Lagrangian density of the present model as follows:

$$\mathcal{L}\left[R(t), \dot{R}(t)\right] = \frac{6}{c^2}\left(\frac{\dot{R}^2 - c^2}{R^2} + \frac{\Lambda}{3}\right), \qquad (24.16)$$

where we have eliminated the term including the second-order time derivative, which should not appear in the Lagrangian density, by reducing it to the total time differential. Substituting (24.16) into (24.15), we obtain

$$I_G = \int_t dt L(t) = -\frac{3\pi c^2}{4G} \int_t dt \left(R\dot{R}^2 - Rc^2 + \frac{\Lambda}{3}R^3\right), \qquad (24.17)$$

where the Lagrangian $L(t)$ is given by

$$L(t) = -\frac{3\pi c^2}{4G}\left(R\dot{R}^2 - Rc^2 + \frac{\Lambda}{3}R^3\right).$$ (24.18)

The Lagrange equation for the scale radius $R(t)$ is then given by

$$\frac{\partial L}{\partial R} - \frac{d}{dt}\left(\frac{\partial L}{\partial \dot{R}}\right) = 0.$$ (24.19)

Substituting (24.18) into (24.19), we obtain

$$-\frac{3\pi c^2}{4G}\left(\dot{R}^2 - c^2 + \Lambda R^2 - 2R\ddot{R} - 2\dot{R}^2\right),$$ (24.20)

or

$$2\frac{\ddot{R}}{R} + \frac{\dot{R}^2 + c^2}{R^2} = \Lambda.$$ (24.21)

Thus the Lagrangian (24.18) gives the correct Equation (24.5) for the scale radius $R(t)$. Let us now define the momentum conjugate to the scale radius $R(t)$ as

$$p_R = \frac{\partial L}{\partial \dot{R}} = -\frac{3\pi c^2}{2G}R\dot{R}.$$ (24.22)

Using (24.22) we may express $\dot{R}$ in terms of p_R as follows:

$$\dot{R} = -\frac{2G}{3\pi c^2}\frac{p_R}{R}.$$ (24.23)

Using Equations (24.22) and (24.23) we obtain the Hamiltonian of the present model as follows:

$$H = p_R\dot{R} - L = -\frac{G}{3\pi c^2}\frac{p_R^2}{R} + \frac{3\pi c^2}{4G}\left(Rc^2 - \frac{\Lambda}{3}R^3\right).$$ (24.24)

The Hamiltonian (24.24) is a classical Hamiltonian of the present mini-superspace model. The quantum-mechanical Hamiltonian operator $\hat{H}$ is obtained using the canonical quantization prescription

$$p_R \rightarrow i\hbar\frac{\partial}{\partial R}, \quad i = \sqrt{-1}.$$ (24.25)

The wave equation of the universe in the present mini-superspace model, known as the *Wheeler–DeWitt equation*, is then obtained from the condition

$H = 0$ as follows:

$$\hat{H}\Psi(R) = \left[\frac{\hbar^2 G}{3\pi c^2 R}\frac{\partial^2}{\partial R^2} + \frac{3\pi c^2}{4G}\left(Rc^2 - \frac{\Lambda}{3}R^3\right)\right]\Psi(R) = 0, \quad (24.26)$$

or

$$\left[-\hbar^2\frac{\partial^2}{\partial R^2} + U(R)\right]\Psi(R) = 0, \quad (24.27)$$

with the potential $U(R)$ defined by

$$U(R) = \left(\frac{3\pi c^3 R_0}{2G}\right)^2\left[\left(\frac{R}{R_0}\right)^2 - \left(\frac{R}{R_0}\right)^4\right]. \quad (24.28)$$

The Wheeler–DeWitt equation written as in (24.27) has the familiar form of the one-dimensional Schrödinger equation for a particle with zero total energy moving in the potential $U(R)$. The form of the potential (24.28) clearly indicates that there is a classically forbidden region for $0 < R < R_0$ and a classically allowed region for $R > R_0$. The point $R = R_0$ is a turning point of the potential (24.28). Classically speaking and using the particle analogy, we conclude that a particle at $R = 0$ is stuck there and cannot travel to the region $R > R_0$. However, in quantum mechanics there is a finite probability that the particle can tunnel through the barrier and emerge at the classical turning point $R = R_0$. Thereafter it can evolve classically in the region $R > R_0$.

24.3   The Wave Function of the Universe

In order to find the solutions of the Wheeler–DeWitt equation for the wave function of the universe $\Psi(R)$ and to calculate the tunneling probability of the universe through the potential barrier (24.28), we restructure Equation (24.27) as follows:

$$\frac{d^2\Psi}{dz^2} + Q^2(z)\Psi = 0, \quad (24.29)$$

where $z = R/R_0$ is a dimensionless scale variable and the function $Q^2(z)$ is defined by

$$Q^2(z) = a^2\left(z^4 - z^2\right), \quad a = \frac{3\pi c^3 R_0^2}{2\hbar G} = \frac{9\pi c^5}{2\hbar G\Lambda}. \quad (24.30)$$

Using the Wentzel–Kramers–Brillouin (WKB) approximation, the wave function of the universe $\Psi(z)$ is given by

$$\Psi(z) = \begin{cases} \frac{1}{2}N_0|Q(z)|^{-1/2}\exp|w(z)|, & z < 1 \quad (R < R_0) \\ N_0|Q(z)|^{-1/2}\cos\left(|w(z)| - \frac{\pi}{4}\right), & z > 1 \quad (R > R_0) \end{cases} \tag{24.31}$$

where N_0 is the WKB normalization constant, which is not essential for the present analysis. The function $w(z)$ is given by

$$w(z) = \int_z Q(z)dz = \frac{a}{3}\left(z^2 - 1\right)^{3/2}. \tag{24.32}$$

For example, in the region where $R \gg R_0$ the wave function (24.31) becomes

$$\Psi(z) = N_0 a^{-1/2}\frac{1}{z}\cos\left(\frac{a}{3}z^3 - \frac{\pi}{4}\right). \tag{24.33}$$

and it is a slowly falling periodic function. Let us now turn to the tunneling probability for the universe to make a transition from $R = 0$ to the transition point $R = R_0$. The most general WKB result for the tunneling probability is given by

$$T = \left[1 + \exp\left(2\int_0^1 Q(z)dz\right)\right]^{-1} = \left[1 + \exp\left(-\frac{2a}{3}\right)\right]^{-1}. \tag{24.34}$$

Substituting a from the result (24.30) into (24.34), we finally obtain

$$T = \left[1 + \exp\left(-\frac{3\pi c^5}{\hbar G\Lambda}\right)\right]^{-1}. \tag{24.35}$$

It should be noted that in many quantum-mechanical problems the exponential in the expression for the tunneling probability (24.34) is a large quantity. It is therefore customary in the literature to use the approximate expression

$$T = \left[1 + \exp\left(2\int_0^1 Q(z)dz\right)\right]^{-1} \approx \exp\left(-2\int_0^1 Q(z)dz\right). \tag{24.36}$$

However, since the gravitational action integral is a negative quantity, such an approximation is invalid in the present discussion and only the more general result (24.34) can be used. Even though the energy density ρ_{vac} and thereby the cosmological constant Λ of the vacuum-dominated Planck epoch are very difficult to estimate, in order to get the idea of the orders of magnitude of these quantities let us assume that the quantity a given by (24.30) is of the order of unity. This gives the tunneling probability of the order of $T \sim 2/3$. Using the result (24.30), we obtain

$$a = \frac{9\pi c^5}{2\hbar G\Lambda} \sim 1 \Rightarrow \Lambda \sim \frac{9\pi c^5}{2\hbar G} = 4.87 \times 10^{+87}\frac{1}{s^2}. \tag{24.37}$$

Now, using $\Lambda = 8\pi G\rho_{vac}$, we have

$$\rho_{vac} = \frac{\Lambda}{8\pi G} \sim 2.9 \times 10^{+96}\frac{kg}{m^3}. \tag{24.38}$$

The result (24.38) indicates an enormous energy density of the universe in the vacuum-dominated Planck epoch. Furthermore, using the result (24.7) we can estimate the radial distance of the turning point of the potential (24.28) as follows:

$$R_0 = ct_0 = c\sqrt{\frac{3}{\Lambda}} \sim 7.44 \times 10^{-36} \text{ m.} \tag{24.39}$$

The length scale of this order of magnitude is sometimes called the Planck length. The time scale t_0 corresponding to this length scale is then given by

$$t_0 = \sqrt{\frac{3}{\Lambda}} \sim 2.48 \times 10^{-44} \text{ s,} \tag{24.40}$$

and it is roughly $t_0 \leq 10^{-43}$ s, which is in agreement with our assertion in the introduction to the present chapter.

A remarkable, and surprising, property of the wave function of the universe is the fact that it is time independent and depends only on the space-time geometry and the matter field content. A possible interpretation of the wave function $\Psi(R)$ is that it measures probabilistic correlations between the geometry and the matter field content. In this interpretation it is possible to use some function of the matter fields, e.g., the energy density of the matter fields, as a surrogate time variable. However, neither the role of time in the quantum cosmology nor the interpretation of the wave function of the universe is fully understood yet.

We also note that just having the wave equation of the universe does not resolve the issue of the quantum evolution of the universe. To be more specific about the quantum evolution of the universe, we need information about the initial quantum state. There are currently several proposals for such an initial quantum state giving different physical predictions. If such predictions can be made to be sufficiently precise, then the present-day observations could be used, at least in principle, to test these different proposals for the initial quantum state of the early universe.

The quantum cosmology efforts around the world today are very ambitious and the very limited presentation given in the present chapter only gives some flavor of this exciting subject. It is our hope that after mastering the material discussed in this chapter, an interested reader will be able to proceed to more advanced monographs on the subject. It is also important to note that despite a number of speculative efforts to create a self-consistent and empirically supported cosmological quantum theory, a lot of work remains to fulfill that goal, and several fundamental questions still remain unresolved.

Bibliography

[1] M. Berry, *Principles of Cosmology and Gravitation*. Adam Hilger, Bristol, 1989.

[2] M. Carmeli, *Classical Fields: General Relativity and Gauge Theory*. Wiley, New York, 1982.

[3] N. Dalarsson, "Phase-Integral Approach to the Wave Equation of the Universe." *Modern Physics Letters A* Vol. 10, No. 33, 1995.

[4] A. Einstein, "Die Grundlage der allgemeinen Relativitätstheorie." *Annalen der Physik* Vol. 49, 1916.

[5] J. Foster and J. D. Nightingale, *A Short Course in General Relativity*. Longman Scientific and Technical, Harlow Essex, 1979.

[6] E. W. Kolb and M. S. Turner, *The Early Universe*. Addison-Wesley, Reading, Massachusetts, 1990.

[7] C. Moller, *The Theory of Relativity*. Oxford University Press, London, 1952.

[8] L. D. Landau and E. M. Lifshitz, *The Classical Theory of Fields*. Addison-Wesley, Reading, Massachusetts, 1959.

[9] W. Panofsky and M. Phillips, *Classical Electricity and Magnetism*. Addison-Wesley, Reading, Massachusetts, 1955.

[10] W. Rindler, *Essential Relativity: Special, General and Cosmological*. Springer-Verlag, Berlin, 1977.

[11] J. Stewart, *Advanced General Relativity*. Cambridge University Press, Cambridge, U.K., 1991.

[12] N. Straumann, *General Relativity and Relativistic Astrophysics*. Springer-Verlag, Berlin, 1981.

[13] R. C. Tolman, *Relativity, Thermodynamics and Cosmology*. Clarendon Press, Oxford, 1934.

[14] R. M. Wald, *General Relativity*. The University of Chicago Press, Chicago, 1984.

[15] S. Weinberg, *Gravitation and Cosmology*. Wiley, New York, 1972.

Index

A

Absolute derivatives,
    of vectors, 69
    of tensors, 69–70
Action,
    integral, 90
    of electromagnetic field, 151
Arc length,
    of curves, 48–9
    element, 48
Associated tensors,
    vectors, 46–7
    tensors, 46–8

B

Black holes,
    mini, 222
    region, 206
    supermassive, 222

C

Christoffel symbols,
    of the first kind, 63
    of the second kind, 63
Conservation law of,
    energy–momentum, 130
    angular momentum, 131
    electric charge, 149
    mass, 185

Continuity equation,
    of electric charge, 149
    of mass, 185
Coordinates,
    of systems, 6
    Descartes, 18–19
    orthogonal, 51
    physical, 52–3
    spherical, 19
Cosmic,
    background radiation, 225
    dynamics, 239–52
    fluid, 227
    time, 234
Cosmological,
    principle, 227
    red shift, 235–7
    scale, 225
Curvature,
    constant, 233
Curvature tensor,
    alternative expression, 102
    covariant, 101
    cyclic identity, 101
    mixed, 101

D

Deceleration,
    parameter, 237
Density,
    charge, 148
    critical (closure), 254

current, 149
Lagrangian, 149

E

Eigenvectors,
  eigenvectors, 55
  orthonormal, 57
Einstein,
  –de Sitter solution, 255
  field equations, 188–91
  tensor, 239–50
Electromagnetic,
  current vector, 147–9
  field invariants, 145
  field tensor, 137
Energy,
  momentum, 126
  rest energy, 127
  total energy, 126
Energy–momentum tensor,
  of matter distribution, 187
Equation of motion,
  of a free particle, 124
  temporal, 216
Epoch,
  matter-dominated, 252
  Planck, 266
  radiation-dominated, 266
  vacuum-dominated, 266–7
Events,
  interval between, 112–16
  horizon, 219

F

Flux of,
  electric charge, 148
  mass, 185
Friedmann,
  equations, 239, 250–2

G

Gauss,
  theorem, 85, 86–7
  curvature, 106

Gauge,
  condition, 141
  of the theory, 141
Geodesic,
  lines, 89–95
  equation, 89, 92–5
Geometry,
  pseudo-Euclidean, 166
  pseudo-Riemannian, 166
Gravitational,
  constant, 173
  field, 165–76
  field potentials, 166
  potential, 171

H

Hamiltonian,
  Δ-operator, 79
  operator, 269
  variational principle, 90
Hubble,
  constant, 226
  law, 225, 226

I

Indices,
  dummy indices, 10
  lower indices, 5–6
  upper indices, 5–6
Inertial systems,
  local, 167
  of reference, 111
Interaction,
  gravitational, 165
  maximum speed of, 112

J

Jacobian,
  of transformation, 33
  Jacobi theorem, 87

K

Kinetic,
  energy, 128
  energy–momentum tensor, 186

L

Lagrangian,
  equation, 89–92
  function, 90
  of charged particle, 136
Linear operators,
  eigenvalues of, 56
  main directions of, 58
  secular equation of, 56
Light,
  cones, 220
  speed of, 112
Lorentz,
  condition, 141
  force, 137
  gauge, 141
  transformation, 116–19

M

Matrix,
  adjunct matrix, 45
  cofactors of, 45
  column matrix, 5
  determinant of, 13
  transposed matrix, 45
  unit matrix, 45
Metric,
  metric, 18
  Schwarzschild, 204, 205
  Robertson–Walker, 225–37
Metric spaces,
  Euclidean, 18
  pseudo-Euclidean, 22
  pseudo-Riemannian, 22
  Riemannian, 18–22

N

Newtonian,
  dust, 260
  Newton law, 193–5

nonrelativistic limit, 173
  universe, 260
Nonrelativistic limit,
  of action integral, 125
  of kinetic energy, 128
  of Lagrangian, 125
  of momentum, 128
  Newtonian, 173

O

Operators,
  commutator of, 103
  operator expression, 103

P

Perihelion,
  advance, 207–15
  of a planet, 207
  shift, 215
Potential,
  electromagnetic, 135
  electric scalar, 135
  magnetic vector, 135

Q

Quantum,
  cosmology, 265–73
  theory, 265

R

Ricci,
  scalar, 105
  tensor, 104

S

Scale,
  radius, 234
Scalars,
  pseudoscalars, 42
  scalar product, 18

Space-time,
  coordinates, 114
  geometry, 165
  metric of, 165
Systems,
  first-order system, 6
  of reference, 111
  second-order system, 6
  third-order system, 7
  arbitrary-order system, 7–8

T

Tensors,
  tensor capacity, 36
  tensor density, 35
Transformation,
  Gallilei, 118
  law, 23
  of coordinates, 23
  of energy–momentum, 128–30

U

Unit systems,
  δ-symbol, 11
  e-symbol, 12
Universe,
  dark matter in, 259
  expansion of, 225
  large-scale structure, 225
  lifespan of, 257
  matter-dominated, 252
  nonstatic model of, 239
  radiation-dominated, 252
  wave function of, 270–3

V

Vectors,
  axial vectors, 42
  curl vector, 86
  of electric field, 137
  of magnetic induction, 137
  polar vectors, 42
  surface vector, 86
  tangent unit vectors, 49
  unit vectors, 49

W

Wheeler–De Witt,
  equation, 266–70
World,
  horizon, 226
  line, 113
  point, 112